ORDER NUMBER EA-AC 60-14

AVIATION INSTRUCTOR'S HANDBOOK

Revised 1977

U.S. DEPARTMENT OF TRANSPORTATION
Federal Aviation Administration

International Standard Book Number 0-89100-308-8
For sale by: IAP, Inc.
Mail to: P.O. Box 10000, Casper, WY 82602-1000
Ship to: 7383 6WN Road, Casper, WY 82604-1835
(800) 443-9250 • (307) 266-3838 • FAX: 307-472-5106

IAP, Inc.
7383 6WN Road, Casper, WY 82604-1835

PREFACE

The term "Aviation Instructor," as used in this handbook, includes the flight instructor, the ground instructor, the aircraft maintenance instructor, and the instructor of air crewmembers. These instructors must understand the basic principles and processes of learning and teaching if they are to attain professional competence. The primary objective of this handbook is, therefore, to provide aviation instructors with comprehensive, accurate, and easily understood information on learning and teaching, and to relate this information to their task of conveying aeronautical knowledge and skill to students.

This handbook, developed at the FAA Aeronautical Center by members of the Flight Standards Service and published as Advisory Circular 60–14, supersedes and cancels the FAA Flight Instructor's Handbook, AC 61–16A. Section One of this handbook contains general information on learning, teaching, critiquing, and evaluating; Section Two deals with flight instructor activities; and Section Three is concerned with aircraft maintenance instructor responsibilities.

Flight instructors should use this handbook in conjunction with the FAA Flight Training Handbook, AC 61–21, which provides detailed descriptions and analyses of flight training maneuvers. Aircraft maintenance instructors should use this handbook in conjunction with the FAA Airframe and Powerplant Mechanics General Handbook, the FAA Airframe and Powerplant Mechanics Powerplant Handbook, and the FAA Airframe and Powerplant Mechanics Airframe Handbook.

Some of the material in this handbook has been derived from the United States Air Force Manual, 50–9, *Principles and Techniques of Instruction*, and the Flight Instructor Refresher Course developed by the Department of Aviation, Ohio State University, under a grant from the Link Foundation. Other material has been derived from the *Aviation Psychology Manual for Flight Instructors*, prepared by Mr. Harold Holmes of the National Safety Council and by Mr. Thomas Hogan, for use in courses given by the Chicago Teachers College.

We gratefully acknowledge the valuable assistance of these organizations and individuals and appreciate their kind permission in allowing us to draw from their works.

Comments regarding this handbook may be directed to U.S. Department of Transportation, Federal Aviation Administration, Flight Standards National Field Office, AFS–500, P.O. Box 25082, Oklahoma City, Oklahoma 73125.

CONTENTS

Contents—*Continued*

SECTION ONE
GENERAL

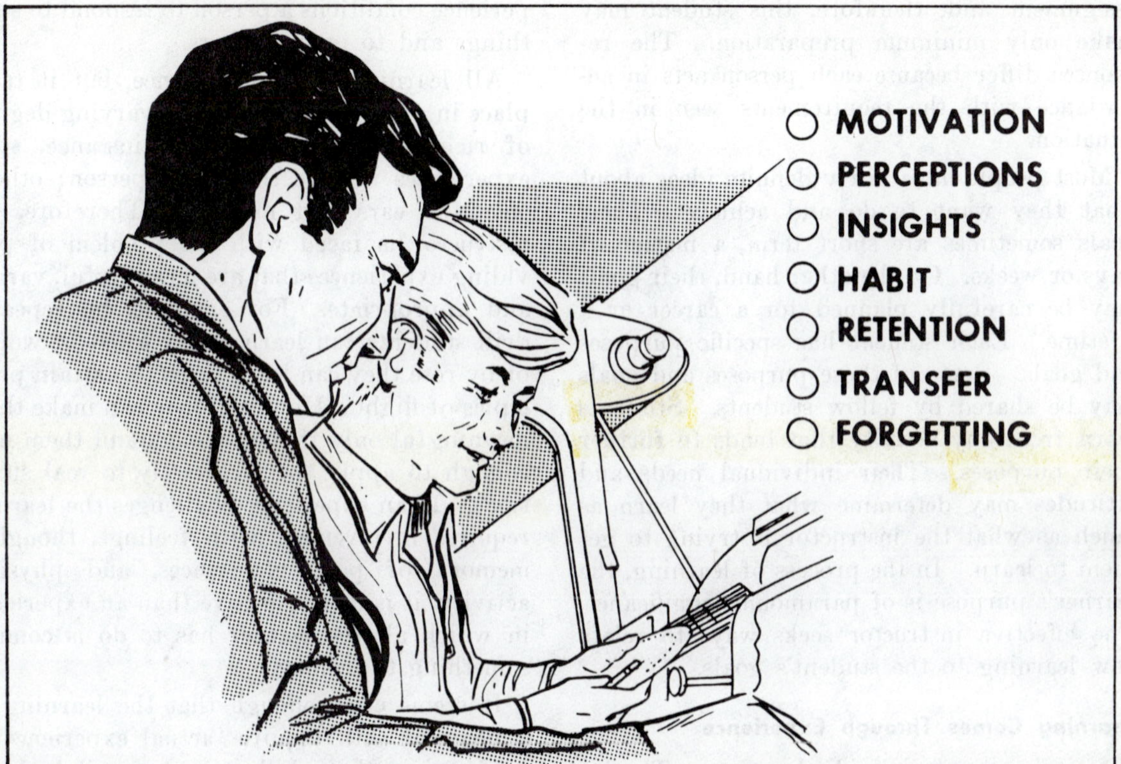

CHAPTER I. THE LEARNING PROCESS

The expert aviation instructor is master of many skills and fields of knowledge. *What* is taught demands technical competence in these areas; but *how* the teaching is accomplished depends largely on the instructor's understanding of how people learn and the ability to apply that understanding. In large measure this handbook could be viewed as a study of applied educational psychology, for the subject underlies virtually everything with which the instructor is concerned. In this chapter, however, only that branch of psychology dealing directly with learning is considered.

DEFINITION OF LEARNING

The ability to learn is one of humanity's most outstanding characteristics. Learning occurs continuously throughout a person's lifetime. To define learning, it is necessary to analyze what happens to the individual. As a result of a learning experience, an individual's way of perceiving, thinking, feeling, and doing may change. Thus, learning can be defined as *a change in behavior as a result of experience.* The behavior can be physical and overt, or it can be intellectual or attitudinal, not easily seen. Psychologists generally agree, however, on some characteristics of learning. The instructor should understand these and turn them to good use.

CHARACTERISTICS OF LEARNING

Learning is Purposeful

Each student sees a learning situation from a different viewpoint. Each student is a unique individual whose past experience affects readiness to learn and understanding of the requirements involved. For example, an instructor may give two student aircraft maintenance technicians the assignment of learning certain inspection procedures. One student may thoroughly learn and be able to competently present the assigned material. Because of aviation background and future goals, that student realizes the value of, and the need for, learning the procedures. A second student's goal may be to merely comply with the instructor's assignment and, therefore, this student may make only minimum preparation. The responses differ because each person acts in accordance with the requirements seen in the situation.

Most people have fairly definite ideas about what they want to do and achieve. Their goals sometimes are short term, a matter of days or weeks. On the other hand, their goals may be carefully planned for a career or a lifetime. Each student has specific purposes and goals. Some of those purposes and goals may be shared by fellow students. Students learn from any activity that tends to further their purposes. Their individual needs and attitudes may determine what they learn as much as what the instructor is trying to get them to learn. In the process of learning, the learner's purpose is of paramount significance. The effective instructor seeks ways to relate new learning to the student's goals.

Learning Comes Through Experience

Learning is an individual process. The instructor cannot do it for the student; knowledge cannot be poured into the student's head. The student can learn only from individual experiences.

LEARNING CANNOT BE ACHIEVED IN THIS MANNER

"Learning" and "knowledge" cannot exist apart from a person. A person's knowledge is a result of experience, and no two people have had identical experiences. Even when observing the same event, two people react differently; they learn different things from it, according to the manner in which the situation affects their individual needs. Previous experience conditions a person to respond to some things and to ignore others.

All learning is by experience, but it takes place in different forms and in varying degrees of richness and depth. For instance, some experiences involve the whole person; others, only the ears and memory. Therefore, the instructor is faced with the problem of providing experiences that are meaningful, varied, and appropriate. For example, by repeated drill students can learn to say a list of words, or by rote they can learn to recite certain principles of flight. However, they can make them meaningful only if they understand them well enough to apply them correctly to real situations. If an experience challenges the learner, requires involvement with feelings, thoughts, memory of past experiences, and physical activity, it is more effective than an experience in which all the learner has to do is commit something to memory.

It seems clear enough that the learning of a physical skill requires actual experience in performing that skill. Student pilots learn to fly airplanes only if their experiences include flying them; student aircraft maintenance technicians learn to overhaul powerplants only by actually performing that task. Mental habits are also learned through practice. If students are to use sound judgment and solve problems well, they must have had learning experiences in which they have exercised judgment and applied their knowledge of general principles in the solving of realistic problems.

Learning is Multifaceted

If instructors see their objective as being only to train their students' memory and muscles, they under-estimate the potential of the teaching situation. Students may have learned much that the instructor had not intended, for they did not leave their thinking minds or feelings at home, just because these were not included in the instructor's plan.

Psychologists sometimes classify learning by types: verbal, conceptual, perceptual, motor, problem solving, and emotional. However useful these divisions may be, they are artificial. For example, a class learning to apply the scientific method of problem solving may learn the method by trying to solve real problems.

But in doing so, it also engages in "verbal learning" and "sensory preception" at the same time. Each student approaches the task with preconceived ideas and feelings, and for many students these ideas change as a result of experience. The learning process, therefore, may include verbal elements, conceptual elements, perceptual elements, emotional elements, and elements of problem solving, all taking place at once.

Learning is multifaceted in still another sense. While learning the subject at hand, students may be learning other things as well. They may be developing attitudes about aviation—good or bad, depending on what they experience. Under a skillful instructor, they may learn self-reliance. The list is seemingly endless. This learning is sometimes called "incidental," but it may have a great impact on the total development of the student.

Learning is an Active Process

Students do not soak up knowledge like a sponge absorbs water. The instructor cannot assume that students remember something just because they were present in the classroom, shop, or airplane when the instructor "taught" it. Neither can the instructor assume that the students can apply what they know because they can quote the correct answer from the book. For the students to learn, they must react and respond, perhaps outwardly, perhaps only inwardly, emotionally, or intellectually. But if learning is a process of changing behavior, clearly that process must be an active one.

LAWS OF LEARNING

One of the pioneers in educational psychology was Professor Edward L. Thorndike, Teachers College, Columbia University, New York. Early in this century Professor Thorndike postulated several "laws" of learning. These rules or principles seemed generally applicable to the learning process. In the years since, other psychologists have found that learning is a more complex process than some of these "laws" suggest. While Professor Thorndike's laws seem to have significant exceptions, they still provide an insight into the learning process and are included in this chapter for that reason.

The "laws" that follow are not necessarily as Professor Thorndike stated them. During the years they have been restated and supplemented, but in essence, they may be attributed to him. The first three are the basic laws, as originally identified: the *law of readiness*, the *law of exercise*, and the still generally accepted *law of effect*. The following three laws were added later as a result of experimental studies: the *law of primacy*, the *law of intensity*, and the *law of recency*.

Law of Readiness

Individuals learn best when they are ready to learn, and they do not learn much if they see no reason for learning. Getting students ready to learn is usually the instructor's responsibility. If students have a strong purpose, a clear objective, and a well-fixed reason for learning something, they make more progress than if they lack motivation. Readiness implies a degree of single-mindedness and eagerness. When students are ready to learn, they meet the instructor at least halfway, and this simplifies the instructor's job.

Under certain circumstances, the instructor can do little, if anything, to inspire in students a readiness to learn. If outside responsibilities, interests, or worries weigh too heavily on their minds, if their schedules are overcrowded, or if their personal problems seem insoluble, students may have little interest in learning.

Law of Exercise

This law states that those things most often repeated are best remembered. It is the basis of practice and drill. The human memory is not infallible. The mind can rarely retain, evaluate, and apply new concepts or practices after a single exposure. Students do not learn to weld during one shop period or to perform cross-wind landings during one instructional flight. They learn by applying what they have been told and shown. Every time practice occurs, learning continues. The instructor must provide opportunities for students to practice or repeat and must see that this process is directed toward a goal.

Law of Effect

This law is based on the emotional reaction of the learner. It states that learning is strengthened when accompanied by a pleasant or satisfying feeling, and that learning is weakened when associated with an unpleasant feeling. An experience that produces feelings of defeat, frustration, anger, confusion, or futility are unpleasant for the student. If, for example, an instructor attempts to teach landings during the first flight, the student is likely to feel inferior and be dissatisfied.

Instructors should be cautious. Impressing students with the difficulty of an aircraft maintenance problem, flight maneuver, or flight crewmember duty can make the teaching task difficult. Usually it is better to tell students that a problem or maneuver, although difficult, is within their capability to understand or perform. Whatever the learning situation, it should contain elements that affect the students positively and give them a feeling of satisfaction.

Law of Primacy

Primacy, the state of being first, often creates a strong, almost unshakable, impression. For the instructor, this means that what is taught must be right the first time. For the student, it means that learning must be right. "Unteaching" is more difficult than teaching. If, for example, a student aircraft maintenance technician learns a faulty riveting technique, the instructor will have a difficult task in unteaching the bad habits and reteaching correct ones. Every student should be started right. The first experience should be positive and functional and lay the foundation for all that is to follow.

Law of Intensity

A vivid, dramatic, or exciting learning experience teaches more than a routine or boring experience. A student is likely to gain greater understanding of stalls by performing them than from merely reading about them. The law of intensity, then, implies that a student will learn more from the real thing than from a substitute. In contrast to flight instruction and shop instruction, the classroom imposes limitations on the amount of realism that can be brought into teaching. The instructor should use imagination in approaching reality as closely as possible. Mockups, colored slides, movies, filmstrips, charts, posters, photographs, and other audio-visual aids can add vividness to classroom instruction.

Law of Recency

The things most recently learned are best remembered. Conversely, the further a student is removed time-wise from a new fact or understanding, the more difficult it is to remember it. It is easy, for example, for a student to recall a torque value used a few minutes earlier, but it is usually impossible to remember an unfamiliar one used a week earlier. Instructors recognize the law of recency when they carefully plan a summary for a ground school lesson, a shop period, or a post-flight critique. The instructor repeats, restates, or reemphasizes important matters at the end of a lesson to make sure that the student remembers them. The law of recency often determines the relative positions of lectures within a course of instruction.

HOW PEOPLE LEARN

Perceptions

Initially, all learning comes from perceptions which are directed to the brain by one or more of the five senses (sight, hearing, touch, smell, and taste). Psychologists have determined through experiments that normal individuals acquire about 75% of their knowledge through the sense of sight, 13% through hearing, 6% through touch, 3% through smell, and 3% through taste. They have also found that learning occurs most rapidly when information is received through more than one sense.

MOST KNOWLEDGE IS ACQUIRED THROUGH SIGHT

Perceiving involves more than the reception of stimuli from the five senses. Perceptions result when a person gives meaning to sensations. People base their actions on the way they believe things to be. The experienced aircraft maintenance technician, for example, perceives an engine malfunction quite differently than does an inexperienced student.

Real meaning comes only from within a person, even though the sensations which evoke these meanings result from external stimuli. Because the meaning which is derived from the information furnished by the senses may depend on many factors within each person concerned, and because perceptions are the basis of all learning, a knowledge of the factors which affect the perceptual process is very important to the instructor.

Factors Which Affect Perception

Among the factors which affect an individual's ability to perceive are that person's: (1) *physical organism*, (2) *basic need*, (3) *goals and values*, (4) *self-concept*, (5) *time and opportunity*, and (6) recognition of the *element of threat*.

The *physical organism* is the vehicle by which individuals become aware of, and operate in, the world of which they are a part. Pilots, for example, must be able to see, to hear, to feel, and to respond adequately while they are in the air. A person whose perceptual apparatus distorts reality is denied the right to fly at the time of the first medical examination.

A person's *basic need* is to maintain and enhance the organized self. The self is complete. It is a person's past, present, and future combined; it is both physical and psychological. A person's most fundamental, pressing need is to preserve and perpetuate this self. All perceptions are affected by this need.

Just as the food one eats and the air one breathes become the physical self, so do the sights one sees and the sounds one hears become the pyschological self. Psychologically, we are what we perceive. A person has physical barriers which keep out those things that would be damaging to the physical being, such as blinking at an arc weld or flinching from a hot iron. So likewise, a person has perceptual barriers that block those sights, sounds, and feelings which pose a psychological threat.

Helping people learn, then, requires finding ways to aid them in developing better perceptions in spite of their defense mechanisms. Since a person's basic need is to maintain and enhance the self, the instructor must recognize that anything that is asked of the student which may be interpreted by the student as imperiling this self will be resisted or denied. To teach effectively, it is necessary to work with this life force.

Perceptions depend on one's *goals and values*. Every experience and sensation which is funneled into one's central nervous system is colored by the individual's own beliefs and value structures. Spectators at a ball game may "see" an infraction or foul differently depending on which team they support. The precise kinds of commitments and philosophical outlooks which the student holds are important for the instructor to know, since this knowledge will assist in predicting how the student will interpret experiences and instructions.

Motivations are also a product of one's value structure. Those things which are more highly valued and cherished are pursued; those which are accorded less value and importance are not sought after. Motivations are one of the most important factors in learning. They are affected by many other factors also, and will be discussed in some detail later in this handbook.

Self-concept, how one pictures oneself, is a most powerful determinant in learning. A student's self-image, described in such terms as "confident" and "insecure," has a great influence on the total preceptual process. If a student's experiences tend to support a favorable self-image, the student tends to remain receptive to subsequent experiences. If a learner has negative experiences which tend to contradict self-concept, there is a tendency to reject additional training.

Negative self-concepts inhibit the perceptual processes by introducing pychological barriers which tend to keep the student from perceiving. They may also inhibit the ability to properly implement that which is perceived. That is, they affect unfavorably the "ability to do." Learners who view themselves positively, on the other hand, are less defensive and more ready to "digest" experiences by assimilating all of the instructions and demonstrations offered.

It takes *time and opportunity* to perceive. Learning some things depends on other perceptions which have preceded these learnings, and on the availability of time to sense and relate these new things to the earlier perceptions. Thus, sequence and time are necessary.

A student could probably stall an airplane on the first attempt, regardless of previous experience. Stalls cannot be "learned," however, unless some experience in normal flight has been acquired. Even with such experience, time and practice are needed to relate the new sensations and experiences associated with stalls in order to develop a perception of the stall. In general, lengthening an experience and increasing its frequency are the most obvious ways to faster learning, although this is not always effective. Many factors in addition to the length and frequency of training periods affect the rate of learning. The effectiveness of the use of a properly planned training syllabus is proportional to the consideration it gives the time and opportunity factor in preception.

The *element of threat*. Fear adversely affects students' perception by narrowing their perceptual field. Confronted with threat, students tend to limit their attention to the threatening object or condition. The field of vision is reduced, for example, when an individual is frightened and all the perceptual faculties are focused on the thing that has generated fear.

Flight instruction provides many clear examples of this. During the initial practice of steep turns, a student pilot may focus attention on the altimeter and completely disregard outside visual references. Anything an instructor does that is interpreted as threatening makes the student less able to accept the experience the instructor is trying to provide. It adversely affects all the student's physical, emotional, and mental faculties.

Learning is a psychological problem, not a logical one. Trying to frighten a student through threats of unsatisfactory reports or reprisals may seem logical, but is not effective psychologically. The effective instructor can organize teaching to fit the psychology of the learner. If a situation seems overwhelming, the student feels unable to handle all of the factors involved, and a threat exists. So long as the student feels capable of coping with a situation, each new experience is viewed as a challenge.

Realizing that behavior is a function of the way in which the individual perceives, and knowing that perceptions are affected by any and all of these factors, enables a good instructor to facilitate the learning process by avoiding any actions which prevent the attainment of teaching goals. Teaching is consistently effective only when those factors which influence perceptions are recognized and taken into account.

Insights

Insights involve the grouping of perceptions into meaningful wholes. Evoking these insights is the instructor's major responsibility. To assure that these do occur, it is essential to keep each student constantly receptive to new experiences and to help the student realize the way each piece relates to all other pieces of the total pattern of the task to be learned.

As a simple example, in straight-and-level flight, in an airplane with a fixed-pitch propeller, the RPM will increase when the throttle is opened and decrease when it is closed. On the other hand, RPM changes can also result from changes in airplane pitch attitude without changes in power setting. Obviously, engine speed, power setting, airspeed, and airplane attitude are all related.

Understanding the way in which each of these factors may affect all of the others and knowing the way in which a change in any one of them may affect changes in all of the others is imperative to true learning. This mental relating and grouping of associated perceptions is called insight.

Insights will almost always occur eventually, whether or not instruction is provided. For this reason, it is possible for a person to become a pilot by trial and error, just as one may become a lawyer by "reading law." Instruction, however, speeds this learning process by teaching the relationship of perceptions as they occur, thus promoting the development of insights by the student.

As perceptions increase in number and are assembled by the student into larger "blocks" of learning to become insights, learning becomes more meaningful and more permanent. Forgetting is less of a problem when there are more anchor points to which one can tie insights. It is a major responsibility of the instructor to organize demonstrations and explanations, and to direct student practice, so that the learner has better opportunities to understand the interrelationship of the many kinds of experiences that have been perceived. Pointing out the relationships as they occur, providing a secure and nonthreatening environment in which to learn, and helping the student acquire and maintain a favorable self-concept are most important in fostering the development of insights.

Motivation

Motivation is probably the dominant force which governs the student's progress and ability to learn. Motivations may be negative or positive; they may be tangible or intangible; they may be very subtle and difficult to identify; or they may be obvious.

Negative motivations are those which may engender fears, and be accepted by the student as threats. While they have their uses in certain situations, they are not characteristically as effective in promoting efficient learning as are positive motivations.

Positive motivations are provided by the promise or achievement of rewards. These rewards may be personal or social; they may involve financial gain, satisfaction of the self-concept, or public recognition. Some motivations which can be used to advantage by the instructor include the desire for personal gain, the desire for personal comfort or security, the desire for group approval, and the achievement of a favorable self-image or sense of achievement.

The desire for personal gain, either the acquisition of things or position, is a basic motivation for all human endeavor. An individual may be motivated to dig a ditch or to design a supersonic airplane solely by the desire for financial gain.

Students are like all other workers in wanting a tangible return for their efforts. If such motivation is to be effective, they must believe that their efforts will be suitably rewarded. These rewards must be constantly apparent to the student during instruction, whether they are to be financial, self-interest, or public recognition.

Many lessons with objectives which are not obvious will pay off well during later instruction, but the student may not appreciate this fact. It is important for the instructor to make the student aware of those applications which are not immediately apparent if motivation is to be maintained. Likewise, the devotion of much time and effort to drill and practice on operations which do not directly contribute to competent performance should be avoided.

The desire for personal comfort and security is a motivation which is often inadequately appreciated by instructors. All students want secure, pleasant conditions and states of being. If they recognize that what they are learning may promote this objective, their interest is easier to attract and hold. Insecure and unpleasant training situations retard learning.

Everyone wants to avoid pain and injury. Students are likely to learn actions and operations which they realize may prevent injury or loss of life. This is especially true when the student knows that the ability to make quick decisions or to act correctly in an emergency results from adequate learning.

The attractive features of the activity to be learned can provide a powerful motivation. Students are anxious to learn skills which may be used to advantage. If they can be made to understand that each learning task to which they are directed will be useful in preparing for the activities for which they undertook training, they will be eager to pursue it.

Group approval is a strong motivating force. Every person wants the approval of friends and superiors. Interest can be stimulated and maintained by building on this natural force. Most students enjoy the feeling of belonging to a group and are interested in attaining an accomplishment which will give them prestige among their fellow students.

Every person seeks to establish a favorable self-image. In certain instances, this self-image may be submerged in a feeling of insecurity or despondency. Fortunately, there is within each person engaged in a task, the belief that success is possible under the proper combination of circumstances and good fortune. This belief can be a most powerful motivating force for most students. This motivation can best be fostered by the instructor through the introduction of perceptions which are solidly based on facts previously learned and which are easily recognized by the student as achievements in learning. Each additional block of learning toward the insight to be developed and toward the ultimate goal, contributes to the confirmation within the student of a favorable self-image. As this confirmation progresses and confidence is achieved, advances will be more rapid and motivation will be strengthened as a result.

Positive motivation is essential to true learning. Negative motivations in the form of reproof and threats should be avoided with all but the most overconfident and impulsive students. Slumps in learning are often due to slumps in motivation. Motivation does not remain at a uniformly high level and may be affected by outside influences, such as physical or mental disturbances or inadequate instruction. The instructor must tailor instruction to the maintenance of the highest possible level of motivation and should be alert to detect and counter relapses in motivation.

LEVELS OF LEARNING

THESE ARE NOT STEPS IN LEARNING

Learning may be accomplished at any of several levels. The lowest level, *rote learning*, is the ability to repeat back something which one has been taught, without understanding or being able to apply what has been learned. Progressively higher levels of learning are *understanding* what has been taught, achieving the *skill to apply* what has been learned and to perform correctly, and associating and *correlating* what has been learned with other things previously learned or subsequently encountered.

For example, a flight instructor may tell a beginning student pilot to enter a turn by banking the airplane with aileron control and applying sufficient rudder in the same direction to prevent slipping and skidding. A student who can repeat this instruction has learned by *rote*. This will never be very use-

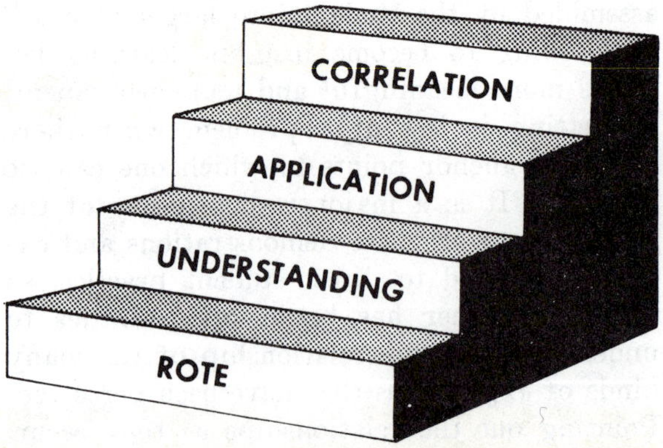

LEVELS OF LEARNING

ful to the student if there is never an opportunity to make a turn in flight or if the student has no knowledge of the function of airplane controls.

With proper instruction on the effect and use of the flight controls, and experience in their use in straight flight, the student can develop these old and new perceptions into an insight on how to make a turn. At this point, the student has developed an *understanding* of the procedure for turning the airplane in flight. This understanding is basic to effective learning, but may not necessarily enable the student to make a correct turn on the first attempt.

When the student understands the procedure for entering a turn, has had turns demonstrated, and has practiced turn entries until consistency has been achieved in an acceptable performance of those entries, the student has developed the *skill to apply* what has been taught. This is a major level of learning, and one at which the instructor is too often willing to stop. Discontinuing instruction on turn entries at this point and directing subsequent instruction exclusively to other elements of piloting performance is characteristic of piecemeal instruction, which is usually inefficient. It violates the "building block" concept of instruction by failing to apply what has been learned to future learning tasks. The "building block" concept will be covered later in more detail.

The highest level of learning, which should be the objective of all instruction, is that level at which the student becomes able to associate an element which has been learned with other segments or "blocks" of learning or accomplishment. The other segments may be items or skills previously learned, or new learning tasks to be undertaken in the future. The student who has achieved this level of learning in turn entries, for example, has developed the *ability to correlate* the elements of turn entries with the performance of such combined and complex piloting operations as those required for the performance of chandelles and lazy eights.

Although the foregoing example deals with the student pilot learning situation, the principles cited apply equally to the student aircraft maintenance technician.

LEARNING SKILLS

Even though the process of learning has many aspects, the main objective or purpose of most instruction is usually the learning of a concept, a generalization, or a skill. The process of learning a skill appears to be much the same, whether it is a motor (physical) or a mental skill. To provide a real illustration of motor learning, please follow the directions below:

> Write the word "learning" 15 times with your left hand (or with your right hand, if you are left handed). Try to improve the speed and quality of your writing.

In the learning task just completed, several principles of motor learning are involved and are discussed in subsequent paragraphs.

Physical Skills Involve More Than Muscles

The above exercise contains a practical example of the multifaceted character of learning. It should be obvious that, while a muscular sequence was being learned, other things were happening as well. The perception changed as the sequence became easier. Concepts of how to perform the skill were developed and attitudes were changed.

Desire to Learn

Thinking back over their past experiences in learning to perform certain skills, students might be surprised at how much more readily they learned those skills that appealed to their own needs (law of readiness). Shorter initial learning time and more rapid progress in improving the skill normally occurred. Conversely, where the desire to learn or improve was missing, little progress was made. A person may read dozens of books a year, but the reading rate will not increase unless there is a deliberate intent to increase it. In the preceding learning exercise, it is unlikely that any improvement occurred unless there was a clear intention to improve. To improve, one must not only recognize mistakes, but also make an effort to correct them. The person who lacks the desire to improve is not likely to make the effort and consequently will continue to practice errors. The skillful instructor relates the lesson objective to the student's intentions and needs and, in so doing, builds on the student's natural enthusiasm.

Patterns to Follow

Logically, the point has been emphasized that the best way to prepare the student to perform a task is to provide a clear, step-by-step example. Having a model to follow permits students to get a clear picture of each step in the sequence—what it is, how to do it. In flight or shop training, the instructor provides the demonstration, emphasizing the steps and techniques. During classroom instruction, an outside expert may be used, either in person or in a film. In any case, however, students need to get a clear impression of what they are to do.

Perform the Skill

Since you have now experienced writing a word with the wrong hand, consider how difficult it would be to tell someone else how to do it. Indeed, even demonstrating how to do it would not result in that person's learning the skill. Obviously, practice is necessary. The student needs coordination between muscles and visual and tactile senses. Learning to perform various aircraft maintenance skills or flight maneuvers requires this sort of practice. There is another benefit of practice. As the student gains proficiency in a skill, verbal instructions mean more. Whereas a long, detailed explanation is confusing before the student begins performing, specific comments are more meaningful and useful after the skill has been partially mastered.

Knowledge of Results

In learning some simple skills, students can discover their own errors quite easily. In learning others, such as complex aircraft maintenance skills, flight maneuvers, or flight crewmember duties, mistakes are not always apparent, or the learner may know that something is wrong but not know how to correct it. In any case, the instructor provides a helpful and often critical function in making certain that the students are aware of their progress. It is perhaps as important for students to know when they are right as when they are wrong. They should be told as soon after the performance as possible, for *they should not be allowed to practice mistakes.* It is more difficult to unlearn a mistake and then learn it correctly, than to learn correctly in the first place. One way to make students aware of their progress is to repeat a demonstration or example and to show them the standard against which they can compare their performance.

Progress Follows a Pattern

The experience of learning to write a word with the wrong hand probably confirmed what has been consistently demonstrated in laboratory experiments on skill learning. The first trials are slow, and coordination is lacking. Mistakes are frequent, but each trial provides clues for improvement in subsequent trials. The learner modifies different aspects of the skill: how to hold the pencil, how to execute finger and hand movement, etc.

Graphs of the progress of skill learning, such as the one shown below, usually follow

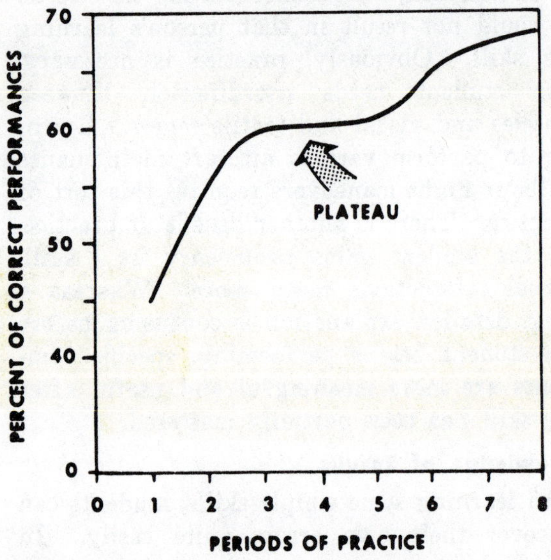

the same pattern. There is rapid improvement in the early trials, then the curve levels off and may stay level for significant periods of effort. Further improvement may seem unlikely. Such a development is a *learning plateau* and may signify any of a number of conditions.

For example, the learner may have reached capability limits; may be consolidating level of skill; interest may have waned; or the learner may need a more efficient method for increasing progress. Keep in mind that the apparent lack of increasing proficiency does not necessarily mean that learning has ceased. The point is that, in learning motor skills, a leveling off process, or a plateau, is normal and should be expected after an initial period of rapid improvement. The instructor should prepare the student for this situation to avert discouragement. If the student is aware of this learning plateau, frustration may be lessened.

Duration and Organization of Lesson

In planning for student performance, a primary consideration is the length of time devoted to practice. A beginning student reaches a point where additional practice is not only unproductive but may even be harmful. When this point is reached, errors increase, and motivation declines. As a student gains experience, longer periods of practice are profitable.

Another consideration is the problem of whether to divide the practice period (and perhaps even the instruction) into segments, or whether to plan on one continuous, integrated sequence. The answer depends on the nature of the skill. Some skills are composed of closely related steps, each dependent on the preceding one; for example, learning to pack a parachute. Other skills are composed of related sub-groups of skills; for example, learning to overhaul an aircraft engine.

Evaluation Versus Critique

If an instructor were to evaluate the fifteenth writing of the word "learning," only limited help could be given toward further improvement. The instructor could judge whether the written word was legible and evaluate it against some criterion or standard, or perhaps even assign it a grade of some sort.

None of these actions would be particularly useful to a beginning student; however, the student could profit by having someone watch the performance and critique it constructively to help eliminate errors.

In the initial stages, practical suggestions are more valuable to the student than a grade. Early evaluation is usually teacher-oriented. It provides a check on teaching effectiveness, it can be used to predict eventual student learning proficiency, and it can help the teacher locate special problem areas. The observations on which the evaluations are based can also identify the student's strengths and weaknesses, a prerequisite for making constructive criticism.

Application of Skill

The final and critical problem is use. Can the student use what has been learned? It is not uncommon to find that students devote weeks and months in school learning new abilities, and then fail to apply these abilities on the job. To solve the problem, two conditions must be present: (1) the student must learn the skill so well that it becomes easy, even habitual to perform it; and (2) the student must recognize the types of situations where it is appropriate to use the skill. This second condition involves the question of transfer of learning, which is discussed later in this chapter.

FORGETTING AND RETENTION

Theories of Forgetting

A consideration of why people forget may point the way to helping them remember. Several theories account for forgetting.

1. *Disuse.* It has long been argued that a person forgets those things which are not used. The high school or college graduate is saddened by the small amount of factual data retained several years after graduation. Since the things which are remembered are those used on the job, a person concludes that forgetting is the result of disuse. But the explanation is not quite so simple. Experimental studies show, for example, that a hypnotized person can describe specific details of an event which would normally be beyond recall. Apparently the memory is there, locked in the recesses of the mind. The difficulty is summoning it up to consciousness.

2. *Interference.* One theory holds that people forget a thing because a certain experience has overshadowed it, or that the learning of similar things has intervened. This theory might explain how the range of experiences after graduation from school causes a person to "lose" knowledge. In other words, new events displace many things that had been learned. From experiments, two conclusions about interference may be drawn: (1) closely similar material seems to interfere with memory more than dissimilar material, and (2) material not well learned suffers most from interference.

3. *Repression.* Freudian psychology advances the view that some forgetting is due to the submersion of ideas into the unconscious mind. Material that is unpleasant or produces anxiety may be treated this way by the individual, but not intentionally. It is subconscious and protective. The repression theory does not appear to account for much forgetfulness of the kind discussed in this chapter, but it does tend to explain some cases.

Retention of Learning

Each of the above theories implies that when a person "forgets" something, it is not actually lost; rather it is unavailable for recall. The instructor's problem, then, is how to make certain that the student's learning is always available for recall. The following suggestions can help.

Teach thoroughly and with meaning. Material thoroughly learned is highly resistant to forgetting. This is suggested by experimental studies; it is also pointed out in the sections on skill learning. Meaningful learning builds patterns of relationship in the student's consciousness. Whereas rote learning is superficial and is not easily retained, meaningful learning goes deep, because it involves principles and concepts anchored in the student's own experience.

The following are five significant principles which are generally accepted as having a direct application to remembering:

1. *Praise stimulates remembering.* Responses which give a pleasurable return tend to be repeated. Absence of praise or recognition tends to discourage one, and any form of

EFFECT

negativism in the acceptance of a response tends to make its recall less likely.

2. *Recall is promoted by association.* Each bit of information or action which is associated with something to be learned tends to facilitate its later recall by the student. Unique or disassociated facts tend to be forgotten unless they are of special interest or application.

3. *Favorable attitudes aid retention.* People learn and remember only what they wish to know. Without motivation there is little chance for recall. The most effective motivations are those based on positive or rewarding objectives.

4. *Learning with all our senses is most effective.* Although we generally receive what we learn through the eyes and ears, other senses also contribute to most perceptions. When several senses respond together, fuller understanding and greater chance of recall is achieved.

5. *Meaningful repetition aids recall.* Each repetition gives the student an opportunity to gain a clearer and more accurate perception of the subject to be learned, but mere repetition does not guarantee retention. Practice gives an opportunity for learning, but does not cause it. Further, it is believed that three or four repetitions provide the maximum effect, after which the rate of learning and probability of retention fall off rapidly.

TRANSFER OF LEARNING

During a learning experience, the student may be aided by things learned previously. On the other hand, it is sometimes apparent that previous learning interferes with the current learning task. Consider the learning of two skills, **A** and **B**. If the learning of **A** helps to learn **B**, *positive transfer* occurs. If learning **A** hinders the learning of **B**, *negative transfer* occurs. For example, the practice of slow flight **A** helps the student learn short field landings **B**; whereas practice in making a landing approach in an airplane **A** may hinder learning to make an approach in a helicopter **B**. It should also be noted that the learning of **B** may affect the retention or proficiency of **A**, either positively or negatively. While these processes may help substantiate the interference theory of forgetting, they are still concerned with the transfer of learning.

It seems clear that some degree of transfer is involved in all learning. That is so because, except for certain inherent responses, all new learning is based upon previously learned experience. People interpret new things in terms of what they already know.

Many aspects of teaching profit by this type of transfer. It may explain why students of apparently equal ability have differing success in certain areas. *Negative transfer* may be hindering the learning of some; *positive transfer* may be helping others. This points to a need to know a student's past experience and what has already been learned. In lesson and curriculum planning, instructors should plan for transfer by organizing course materials and individual lesson materials in meaningful sequence. Each phase should help the students to learn what is to follow.

POSITIVE TRANSFER

The cause of transfer and how it operates has not yet been identified and explained. But no one disputes the fact that transfer does occur. The significance of this ability for the instructor is that the students can be helped to achieve it. The following suggestions are representative of what educational psychologists believe should be done:

1. Plan for transfer as a primary objective. As in all areas of teaching, the chance for success is increased if the teacher deliberately plans to achieve it.
2. Make certain that the students understand that what is learned can be applied to other situations. Prepare them to seek other applications.

3. Assure thorough, high-order learning. Overlearning may even be appropriate. The more thoroughly the students understand the material, the more likely they are to see its relationship to new situations. Avoid rote learning, for it does not foster transfer.
4. Provide meaningful learning experiences that build the students' confidence in their ability to transfer learning. This suggests activities that challenge them to exercise their imagination and ingenuity in applying their knowledge and skills.
5. Use instructional material that helps form valid concepts and generalizations. Use materials that make relationships clear.

HABIT FORMATION

The formation of correct habit patterns from the beginning of any learning process is essential to further learning and for correct performance after the completion of training. It is, therefore, the instructor's responsibility to insist on correct techniques and procedures from the outset of training to provide proper habit patterns. It is much easier to foster proper habits from the beginning of training than to correct faulty ones later.

This is the basic reason for the *building block* technique of instruction, in which each simple task is performed acceptably and correctly before the next learning task is introduced. The introduction of instruction in more advanced and complex operations before the initial instruction has been mastered leads to the development of poor habit patterns in the elements of performance. Faulty performance of the elements are inevitably carried through to all future learning.

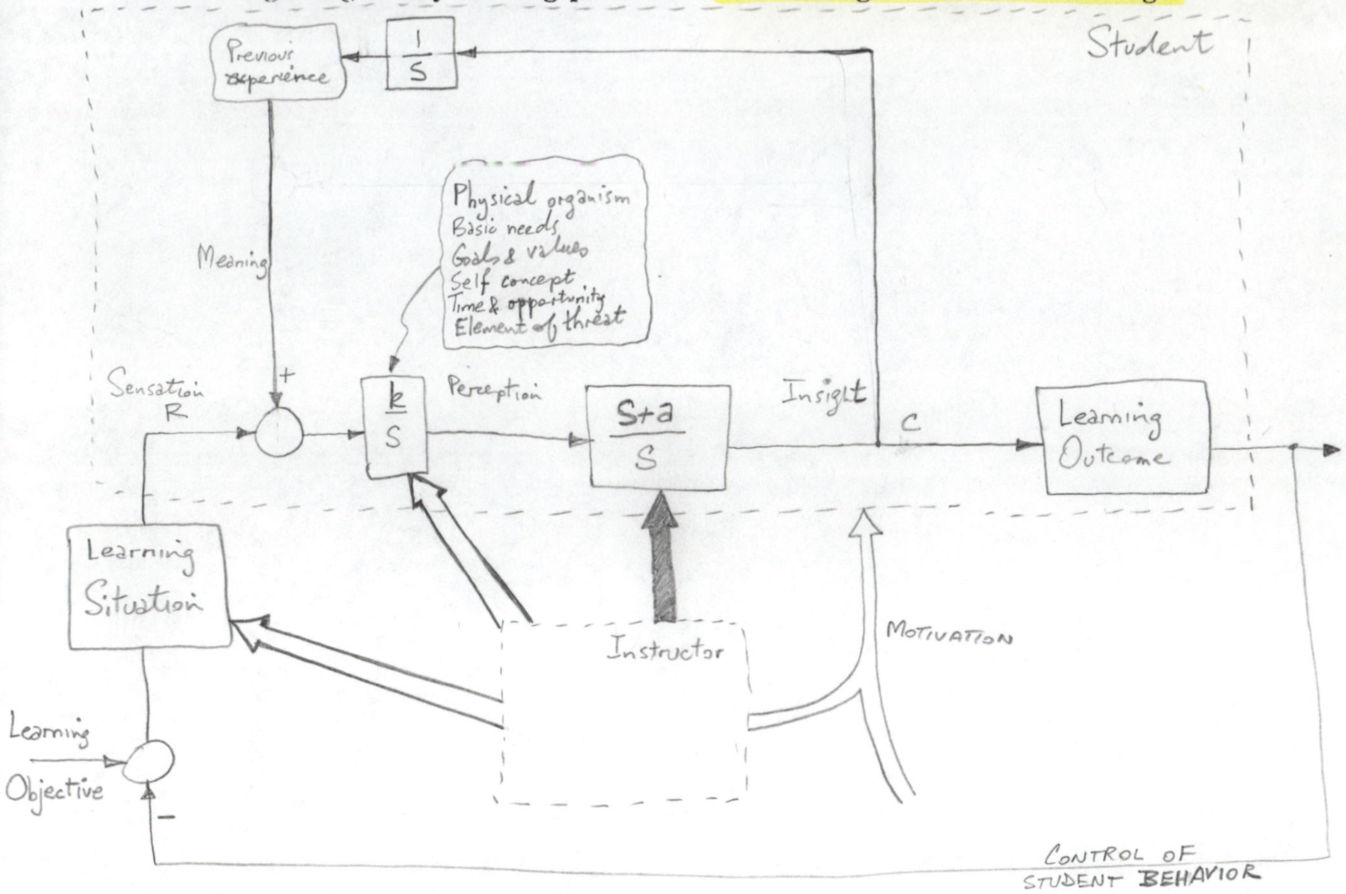

$$\epsilon = R + \frac{C}{S}$$

$$C = \frac{S+a}{S^2} \cdot \epsilon$$

$$= \frac{S+a}{S^2}\left[R + \frac{C}{S}\right]$$

$$C\left[1 - \frac{S+a}{S^3}\right] = \frac{S+a}{S^2} \cdot R$$

The student loop can be unstable in itself:

$$\frac{C}{R} = \frac{G}{1-GH} = \frac{\frac{S+a}{S^2}}{1-\frac{S+a}{S^3}} = \frac{S(S+a)}{S^3 - S - a}$$

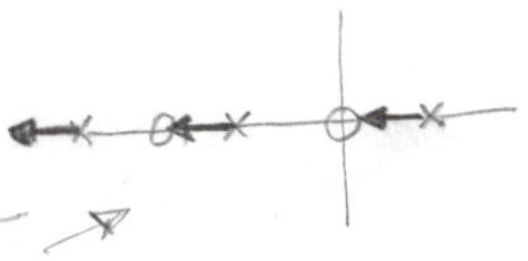

The instructor can stabilize the overall loop and drive the learning outcome towards the learning objective

CHAPTER II. HUMAN BEHAVIOR

By definition, *learning is a change of behavior resulting from experience.* To successfully accomplish the task of helping to bring about this change, the instructor must know why human beings act the way they do. A knowledge of basic human needs and defense mechanisms will aid the instructor in organizing student activities and in promoting a climate conducive to learning.

CONTROL OF HUMAN BEHAVIOR

The relationship between the instructor and the students has a profound impact on how much the students learn. To students, the instructor is a symbol of authority. Students expect the instructor to exercise certain controls, and they recognize and submit to authority as a valid means of control. The instructor's challenge is to know what controls are best for what circumstances. The instructor should create an environment that enables students to help themselves.

Every student works toward a goal of some kind. It may be success itself; it may be a grade or an honor. The instructor directs and controls the behavior of the students and guides them toward a goal. This is a process of directing the students' actions and modifying their behavior. Without the instructor's active intervention, the students would be passive and perhaps resistant to learning. The controls the instructor exercises—how much, how far, to what degree—should be based on more than random selection or trial and error.

See Figure p. 13

Some interesting generalizations have been made about motivated human nature by the late Douglas McGregor, Massachusetts Institute of Technology. While Professor McGregor's assumptions are directed specifically toward industrial management, they have implications for the instructor as well.

1. The expenditure of physical and mental effort in work is as natural as play and rest. The average human being does not inherently dislike work. Depending on conditions, work may be a source of satisfaction and, if so, will be performed voluntarily. On the other hand, when

work is a form of punishment, it may well be avoided, if possible.

2. A human being will exercise self-direction and self-control in the pursuit of goals to which committed.
3. Commitment to goals relates directly to the reward associated with their achievement, the most significant of which is probably the satisfaction of ego.
4. The average human being learns, under proper conditions, not only to accept, but to seek responsibility. Shirking responsibility and lack of ambition are not inherent in human nature. They are usually the consequences of experience.
5. The capacity to exercise a relatively high degree of imagination, ingenuity, and creativity in the solution of common problems is widely, not narrowly, distributed in the population.
6. Under the conditions of modern life, the intellectual potentialities of the average human being are only partially used.

An instructor who accepts McGregor's assumptions begins to see the student as a vast, untapped potentiality. The instructor's ingenuity must be used in discovering how to realize the potentialities of the student. The responsibility rests squarely on the instructor's shoulders. If the student is perceived as lazy, indifferent, unresponsive, uncreative, uncooperative, and antagonistic, McGregor's assumptions imply that the cause lies in the instructor's methods of control. The raw material is there, McGregor maintains and the shaping and directing of it lie in the hands of those who have the responsibility of controlling it.

How to mold a solid, healthy, productive relationship with students depends, of course, on the instructor's knowledge of students as human beings and of the needs, drives, and desires they continually try to satisfy in one way or another. Some of their needs and drives are discussed in subsequent paragraphs.

HUMAN NEEDS

The instructor should always be aware of the fact that the students are human beings. The needs of students, and of all mankind, are given labels by psychologists and are generally organized in a series of levels. The "pyramid of human needs" has been suggested by Abraham Maslow of Brandeis University.

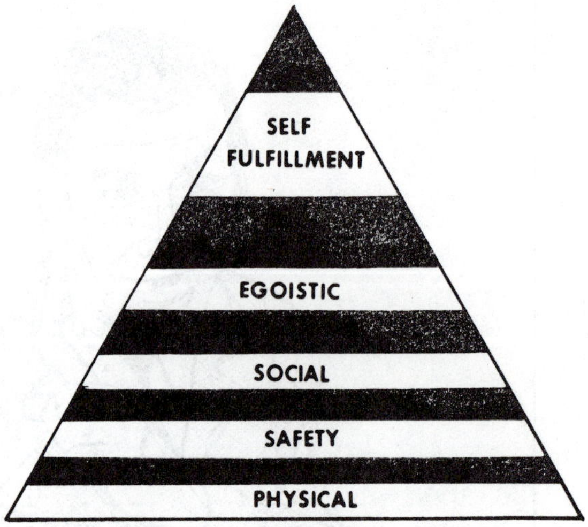

PYRAMID OF HUMAN NEEDS

Physical Needs

At the broadest level are the physical needs. Individuals are first concerned with their need for food, rest, exercise, and protection from the elements. Until these needs are satisfied to a reasonable degree, they cannot concentrate on learning or self-expression. Of course, *once a need is satisfied, it no longer provides motivation*, because a want that is satisfied is no longer a want. Each individual, therefore, strives to satisfy the needs of the next higher level.

Safety Needs

The safety needs are protection against danger, threat, deprivation, and are labeled by some authors as the security needs. Regardless of the label, however, they are real, and student behavior is influenced by them.

Social Needs

If individuals are physically comfortable and have no fear for their safety, their social needs then become the prime influence on their behavior. These needs are to belong, to associate, to give and receive friendship and love. Many studies have demonstrated that a tightly knit, cohesive group, under proper conditions, will be more effective than an equal number of separate individuals. Inasmuch as students are usually separated from normal surroundings, their need for association and for belonging will be more pronounced.

Egoistic Needs

The egoistic needs will usually have a direct influence on the student-instructor relationship. Those needs are two kinds: (1) those that relate to one's self-esteem—needs for self-confidence, for independence, for achievement, for competence, for knowledge; and (2) those needs that relate to one's reputation—needs for status, for recognition, for appreciation, for the deserved respect of one's fellows.

Self-Fulfillment Needs

At the apex of the hierarchy of human needs are those for self-fulfillment, or for realizing one's own potentialities, for continued development, for being creative in the broadest sense of that term. This need of a student should offer the greatest challenge to the instructor. Aiding another in realizing self-fulfillment is perhaps the most worthwhile accomplishment an instructor can achieve.

An instructor should strive to help the student satisfy these needs in a manner that will ensure a healthy environment for learning. When the instructor works for good human relations, the students experience fewer frustrations and can therefore devote more attention to their studies.

DEFENSE MECHANISMS

Certain behavior patterns are called defense mechanisms because they are subconscious defenses against the realities of unpleasant situations. People use these defenses to soften feelings of failure, to alleviate feelings of guilt, and to protect feelings of personal worth and adequacy.

Although defense mechanisms can serve a useful purpose, they can also be hindrances. Because they involve some self-deception and distortion of reality, defense mechanisms do not solve problems. They alleviate symptoms, not causes. Moreover, since they operate on a relatively unconscious level, they are not subject to normal conscious checks and controls. Once an individual realizes there is a conscious reliance on one of these devices, behavior ceases to be a subconscious adjustment mechanism and becomes, instead, an ineffective way of satisfying a need. Common defense mechanisms are *rationalization*, *flight*, *aggression*, and *resignation*.

Rationalization

If students cannot accept the real reasons for their behavior, they may rationalize. This device permits them to substitute excuses for reasons; moreover, they can make those excuses plausible and acceptable to themselves. Rationalization is a subconscious technique for justifying actions that otherwise would be unacceptable. When true rationalization takes place, individuals sincerely believe in their excuses. The excuses seem real and justifiable.

Flight

Students often escape from frustrating situations by taking flight, physical or mental. To take flight physically, students may develop symptoms or ailments that give them satisfactory excuses for removing themselves from frustration. More frequent than physical flights are mental flights, or daydreaming. Mental flight provides a simple and satisfying escape from problems. If students get sufficient satisfaction from daydreaming, they may stop trying to achieve their goals altogether. When carried to extremes, the world of fantasy and the world of reality can become so confused that the dreamer cannot distinguish one from the other.

Aggression

Everyone gets angry occasionally. Anger is a normal, universal human emotion. Angry persons may shout, swear, slam a door, or give in to the heat of emotions in a number of ways. They become aggressive against something or somebody. After a cooling-off period, they may see their actions as childish. In a classroom, shop, or airplane, such extreme behavior is relatively infrequent. Because of social strictures, student aggressiveness is usually subtle. Students may ask irrelevant questions, refuse to participate in the activities of the class, or disrupt activities within their own group. If students cannot deal directly with the cause of their frustration, they may vent their aggressiveness on a neutral object or person not related to the problem.

Resignation

Students may become so frustrated that they lose interest and give up. They may no longer believe it profitable or even possible to work further. They accept defeat. The most obvious and apparent cause for this form of resignation takes place when, after completing the early phase of a course without grasping the fundamentals, a student becomes bewildered and lost in the advanced phase. From that point on, learning is negligible although the student may go through the motions of participating.

THE INSTRUCTOR'S ROLE IN HUMAN RELATIONS

To minimize student frustrations and to help achieve good human relations in the classroom, shop, or during flight training, are basic instructor responsibilities. Instructors can follow several rules which, if adapted to the problem at hand, can be of value.

Keep Students Motivated

Students gain more from wanting to learn than from being forced to learn. All too often students do not realize how a particular lesson or course can help them reach an important goal. When they can see the benefits or purpose of a lesson or course, their enjoyment and their efforts will increase.

Keep Students Informed

Students feel insecure when they do not know what is expected of them or what is going to happen to them. Instructors can minimize such feelings of insecurity by telling students what is expected of them and what they can expect. Instructors can keep students informed in various ways: giving them an overview of the course; keeping them posted on their progress; and giving them adequate notice of examinations, assignments, or other requirements.

Approach Students as Individuals

When instructors limit their thinking to the whole group without considering the individuals who make up that group, their effort is directed at an average personality which really fits no one. Each group has its own personality which stems from the characteristics and interactions of its members. However, each individual within the group has a personality which is unique and which should be constantly considered.

Give Credit When Due

When students do something extremely well, they wish their abilities and efforts to be noticed. Otherwise, they become frustrated. Praise or credit from the instructor is usually ample reward and provides an incentive to do even better. Praise given too freely, however, becomes valueless; but when deserved, it pays dividends in student effort and achievement.

Criticize Constructively

Although it is important to give praise and credit when deserved, it is equally important to identify mistakes and failures. To tell students that they have made errors and not provide explanations does not help them. If a student has made an earnest effort but is told that the work is not satisfactory, with no other explanation, frustration occurs. Errors cannot be corrected if they are not identified, and if they are not identified, they will probably be perpetuated through faulty practice. If, on the other hand, the student is briefed on the errors made and is told how to correct them, progress and accomplishment can be made.

Be Consistent

Students want to please their instructor. This is the same desire that influences much of the behavior of subordinates toward their superiors in industry and business. Naturally, students have a keen interest in knowing what is required to please the instructor. If the same thing is acceptable one day and not acceptable the next, the student becomes confused. The instructor's philosophy and actions must therefore be consistent.

Admit Errors

No one, including the students, expects an instructor to be perfect. The instructor can win the respect of students by honestly acknowledging mistakes. If the instructor tries to cover up or bluff, the students will be quick to sense it. Such behavior tends to destroy student confidence. If in doubt about some point, the instructor should admit it to the students.

These are but a few of the many attitudes and reactions that can help establish good human relations. Good human relations promote more effective learning.

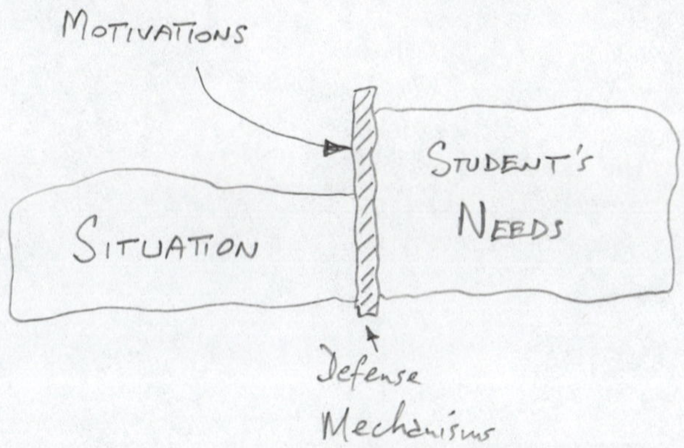

CHAPTER III. EFFECTIVE COMMUNICATION

Communicating as an instructor is an essential but difficult skill. It is difficult because of the variations and complexities in the teaching-learning process. Nevertheless, an examination of the process of communication, by an analysis of its cardinal elements and the significant relationships between them, is useful in gaining a more precise and deeper understanding of the process. Improvement in communication depends, in large measure, on an understanding of the process.

BASIC ELEMENTS OF THE COMMUNICATION PROCESS

Communication takes place when one person transmits ideas or feelings to another person or to a group of people. Its effectiveness is measured by the similarity between the idea transmitted and the idea received.

The process of communication is composed of three elements: (1) *the source* (a sender, speaker, writer, instructor, transmitter, or encoder); (2) *the symbols* used in composing and transmitting the message (words, signs, music); and (3) *the receiver* (a listener, reader, student, or decoder). These elements are dynamically interrelated, and that which affects one influences the others. If a listener has difficulty in understanding the symbols a speaker is using and indicates confusion, the speaker may become puzzled and uncertain, losing selective control of ideas. Thus, communication effectiveness is diminished. On the other hand, when a listener reacts favorably, a speaker is encouraged, and force is added to communication. The relationship between the communicative elements is not only dynamic but also reciprocal. Communication is a complicated, two-way process.

Source

The effectiveness of persons acting in the role of communicators is related to at least three basic factors. *First*, their facility in selecting and using language influences their ability to select symbols that are meaningful to the listeners or readers. *Second*, communicators, consciously or unconsciously, reveal attitudes toward themselves, toward the ideas they are trying to transmit, and toward their receivers. These attitudes must be positive if they are to communicate effectively. Communicators must be confident. They must indicate that they believe their message is important. Communicators must make it clear to their listeners or readers that they believe there is a need to know the ideas presented. *Third*, successful communicators speak or write from a broad background of accurate, up-to-date, stimulating material. Communicators must exercise great care to make certain that they communicate with ideas and feelings that are meaningful to their receivers. Far too often, a speaker or a writer may depend on a highly technical or professional background, with its associated vocabulary, which is meaningful only to others of like background. Reliance on technical language to express ideas to any receiver often impedes effective communication.

Symbols

At its basic level, communication is achieved through the use of simple oral and visual codes. The letters of our alphabet, when translated into words, constitute a basic code. Common gestures and facial expressions form another. But words and gestures are seldom projected in isolation. Ideas are communicated only when symbols are combined in meaningful wholes, in ideas, sentences, paragraphs, speeches, or chapters. Each part of the whole then becomes important for effective communication.

Total ideas must be selected carefully if they are to convey messages which receivers can react to and understand. They must be analyzed to determine which are most suited to starting and concluding the communication, and which ideas can clarify, emphasize, define, limit, and explain—all of which form the basis for effective transmission of ideas from source to receiver. Finally, the development of ideas from simple symbols culminates in the determination of the medium best suited for their transmission. Many channels are available for transmission. Most frequently, communicators select the channels of hearing and seeing. Occasionally, the channel of feeling—of actually touching or manipulating objects—can be used effectively. The most successful communicator, however, uses a variety of channels through which to communicate selected ideas.

Receiver

Effective communicators always remember a basic rule of thumb—*communication succeeds only in relation to the reaction of the receivers.* When the receivers react with understanding, and change their behavior accordingly, then, and only then, has communication taken place.

To understand the process of communication, at least three characteristics of receivers must be understood: their abilities, attitudes, and experiences. *First*, they exercise their ability to question and comprehend the ideas that have been transmitted. Communicators can capitalize on this ability by providing an atmosphere which encourages questioning. Questions may be direct or silent. Readers do read; listeners do listen. Understanding the receiver's abilities is necessary if one is to understand the process of communication. *Second*, the receiver's attitude may be one of resistance, of willingness, or of passive neutrality. Whatever the attitude, communicators must gain the receiver's attention and then retain it. Probably the more they vary their communicative approach, the more successful they will be in this respect. *Third*, the receiver's background, experience, and education frame the target at which communicators must aim. Communicators assume an obligation to assess their receiver's knowledge and to use it as a fundamental guide for the selection and transmittal of ideas. To get the receiver's reaction, the communicator must first reach them. The major barriers to effective communication are usually found in this particular area.

BARRIERS TO EFFECTIVE COMMUNICATION

The nature of language and the ways it is used often lead to misunderstandings. These misunderstandings stem primarily from three barriers to effective communication: the *lack of a common core of experience*, a *confusion between the symbol and the thing symbolized*, and the *overuse of abstractions*.

Lack of Common Core of Experience

Probably the greatest single barrier to effective communication is the lack of a common core of experience between communicator and receiver. Communication can be effective only to the extent that the experiences—physical, mental, or emotional—of the people concerned are similar.

Many people believe that words transport meanings from speaker to listener in the same way that a truck carries bricks from one location to another. But words never carry precisely the same meaning from the mind of the communicator to that of the receiver. In fact, words do not transfer meanings at all. Words, both spoken and written, are merely stimuli that the communicator sets forth to arouse a response of some kind in the receiver. The nature of this response is determined by the receiver's past experiences with the words and the things to which they refer. These experiences give the words their meaning; meaning that is in the mind of the receiver, not in the words themselves. Since a common core of experience is basic to effective communication, a communicator's words cannot communicate meaning to listeners or readers unless they have had some experience with the objects or concepts to which these words refer.

Confusion Between the Symbol and the Thing Symbolized

Words are simply representations. They represent or correspond to anything that exists, that is experienced, or that people talk about. At best, language serves as a map. Just as a useful map accurately represents some specified territory, language should correspond to the objects or concepts that it represents. Like a map that contains errors, a statement that contains inaccuracies implies a relationship that does not exist. Communicators must realize the danger in confusing symbols with the things they symbolize. Effective speakers and writers carefully differentiate between symbols and the things they represent, keeping both in true perspective.

Overuse of Abstractions

Concrete words refer to objects that human beings can experience directly. Abstract words, on the other hand, stand for ideas that cannot be directly experienced, for things that do not call forth mental images in the minds of the receivers. For example, if a communicator is discussing a particular experimental supersonic airplane and refers to it as the X–15, the listeners immediately get a mental image of this airplane. The name X–15 represents a *concrete* reality. It can be seen, touched, and heard. If, however, the communicator says, "I saw an experimental supersonic airplane," the listeners do not form a specific mental image of the X–15 because there are a number of airplanes which fit that description. If the communicator says, "I saw a fast aircraft," a term is being used which is so *abstract* that the listeners cannot form a mental image of the X–15 at all.

Abstract words are necessary and useful. Their purpose is not to bring forth specific items of experience in the minds of receivers, but to serve as shorthand symbols that sum up vast areas of experience. Although abstractions are convenient and useful, they can lead to misunderstanding. The danger in using them is that they will not evoke in a listener's mind the specific items of experience that communicators intend. The receiver has no way of knowing what experiences the speaker or writer intends an abstraction to include. It is common practice to use such abstract terms as "proper measures" and "corrective action." These terms alone fail to convey the communicator's intent. When abstractions are used in communication, they should be linked with specific experiences through examples and illustrations. Even better, the level of abstraction should be reduced by using concrete and specific words as much as possible. By using concrete words, the communicator narrows and gains better control of the image produced in the minds of listeners and readers.

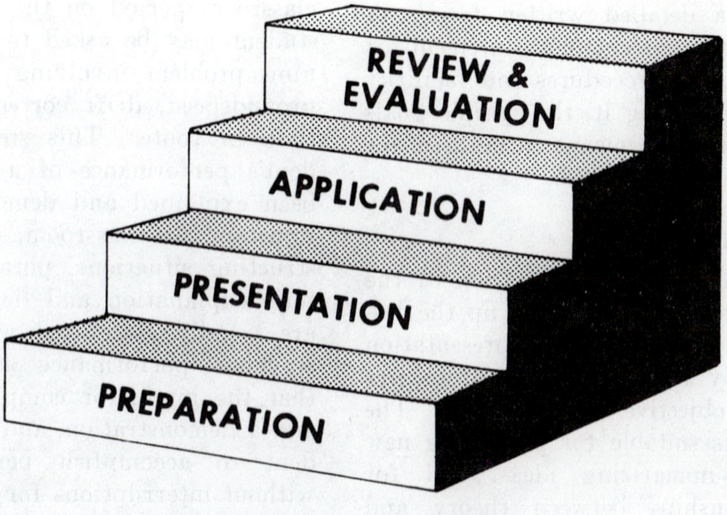

TEACHING STEPS

CHAPTER IV. THE TEACHING PROCESS

Any effective teaching process must be based on the principles of learning which have been discussed in some detail in Chapter I. The learning process does not seem to be naturally divisible into a definite number of steps. Sometimes, it occurs almost instantaneously, as when a child learns about heat from touching a hot stove. In other cases, learning is acquired only through long, patient study and diligent practice.

Although the teaching process *can* be divided into steps, much conflicting material has been written with reference to those steps. Some recognized authorities have specified as few as three steps, while others have broken the teaching process down into seven or eight steps. A close examination of the various lists of steps in the teaching process reveals that different authors are saying essentially the same thing. The chief difference between them is the fact that some authors include only the steps in the actual teaching process, while others include the steps involved in the preparation of the instructor for the job. Another difference is that some authors make separate steps of items like summaries and assignments, while others do not.

The teaching of new material, as reflected in any of the lists, can be broken down into the steps of: (1) *preparation*, (2) *presentation*, (3) *application*, and (4) *review and evaluation*. Discussions in this handbook will center on these four basic steps.

PREPARATION

For each lesson or instructional period, the instructor must determine what is to be covered, the objectives of the lesson, and the goals to be attained. This step should also include home study or other special preparation by the student. As part of the preparation, the instructor should make certain that all necessary supplies, materials, and equipment are readily available and that the equipment is operating properly. The instructor's preparation should include actual reference to the syllabus for the course involved and a study of objectives. It must include the development of a lesson plan if the instructional period is to be effective.

The instructor's lesson plan should be worked out in a detailed written form. A lesson plan is the instructor's statement of lesson objectives, the procedures and facilities to be used in presenting it, the specific goals to be attained, and the means to be used for evaluating the results achieved.

PRESENTATION

This is the instructor's presentation of the knowledge and skills which make up the lesson. The choice of the method of presentation is determined by the nature of the subject matter and the objective in teaching it. The *lecture method* is suitable for presenting new material, for summarizing ideas, and for showing relationships between theory and practice. For example, it is suitable for the presentation of a ground school lesson on aircraft weight and balance. This method is most effective if accompanied by instructional aids and training devices. In the case of a lecture on weight and balance, a chalkboard could be used effectively. The *demonstration-performance method* is desirable for presenting a skill, such as a ground school lesson on the flight computer. In using this method, be sure to tell the facts or demonstrate the steps in the proper order so the students get a clear-cut picture of each separate part of the process or operation.

APPLICATION

This is the student's application of what the instructor presented. In a classroom situation, the student may be asked to explain the new material, or to perform a maneuver or operation. For example, at the end of a classroom period on the flight computer, the student may be asked to work a flight planning problem involving the computation of groundspeed, drift correction, and estimated time en route. This step involves the student's performance of a procedure that has been explained and demonstrated by the instructor. In classroom, shop, and flight instructing situations, portions of the instructor's explanation and demonstration activity are usually alternated with portions of the student's performance activity. It is rare that the instructor completes an explanation and a demonstration, and then allows the student to accomplish performance activities without interruptions for corrections and further demonstrations. It is very important that the student perform the maneuver or operation the right way the first few times, for this is when habits are established. Faulty habits are difficult to correct. After reasonable competence has been attained, the maneuver or operation should be practiced again and again until correct performance becomes almost automatic.

REVIEW AND EVALUATION

This is an integral part of each classroom, shop, or flight lesson. Before the end of the instructional period, the instructor should recapitulate what has been covered during the lesson, and require the students to demonstrate the extent to which the lesson objectives have been met. The instructor's evaluation may be informal and noted only for use in planning the next lesson for the students, or it may be recorded to certify the students' progress in the course.

In either case, the students should be aware of their progress and the advances and deficiencies noted at the conclusion of the lesson. The failure of the instructor to assure that students are cognizant of their progress, or lack of it, may impose a barrier between them. Though it may be slight, it may make further instruction more difficult.

In flight training, aircraft maintenance training, and air crewmember training situations, the instructor must remember that it is rather difficult for students to obtain a clear picture of their progress, since they have little opportunity for a direct comparison with others, especially in the early phases of training. The students recognize that they are in a competitive situation unlike any previously experienced. The unseen competitor is that intangible "competency" which must be achieved. The students' own evaluations can only be subjective. Direct comparisons for them are only possible with the performance of the instructor. Only the instructor can provide a realistic evaluation of performance and progress.

In addition to knowledge and skills learned during the instruction period just completed, each lesson should include a review and evaluation of things previously learned. If the evaluation reveals a deficiency or fault in the knowledge or performances on which the present lesson is predicated, it must be corrected before a new lesson can begin.

If deficiencies or faults not associated with the present lesson are revealed, they should be carefully noted and pointed out. Such corrective measures as are practicable within the limitations of the situation should be taken immediately, but more thorough remedial actions must be included in future lesson plans.

The evaluation of student performance and accomplishment during a lesson should be based on the objectives and goals that were established in the instructor's lesson plan.

CHAPTER V. TEACHING METHODS

Teaching methods may be considered the tools of the instructor's trade. The instructor's skill is determined to a large degree by the ability to organize material and to select and utilize a teaching method appropriate to a particular lesson. Of the various teaching methods in common use, only the *lecture method*, the *guided discussion method*, and the *demonstration-performance method* will be covered in this handbook. There are no definite lines of division between these methods. In a particular situation, an instructor should use more than one method. For example, a good demonstration is usually accompanied by a thorough explanation, which is essentially a lecture.

Because of the constantly increasing emphasis on programmed instruction and since a number of programmed aviation training courses are presently available from industry, a brief description of this method is given at the end of this chapter.

ORGANIZING MATERIAL: LESSON CONTENT

Regardless of the teaching method used (lecture, guided discussion, demonstration-performance, etc.), an instructor must properly organize the material. One effective way to organize a lesson is—introduction, development, and conclusion.

1. Introduction

The introduction should serve several purposes: to establish common ground between the instructor and the students, to capture and hold the attention of the group, to indicate what is to be covered during the presentation and relate this coverage to the entire course, to point out specific benefits the student can expect from the learning, and to establish a receptive attitude toward the subject and lead into the lesson development. In brief, the introduction sets the stage for learning.

a. *Attention.* The instructor might begin by telling a story that relates to the subject and establishes a background for developing learning outcomes. The instructor might gain the students' attention by making an unexpected or surprising statement or by asking a question that helps relate the lesson topic to the welfare of the group. No matter how the instructor introduces the lesson, the main concern should be to gain the students' attention and to focus it on the subject.

b. *Motivation.* The introduction should offer the students specific reasons for needing to be familiar with, to know, to understand, to apply, or to be able to perform whatever they are about to learn. This motivation should appeal to each student personally and accentuate the desire to learn.

c. *Overview.* Every lesson introduction should contain an overview that tells the group what is to be covered during the period. A clear, concise presentation of the objective and the key ideas gives the students a road map of the route to be followed. A good visual aid can help the instructor show the students the route that they are to travel. The introduction should be free of stories, jokes, or incidents that do not help the students focus their attention on the lesson objective. Also, the instructor should avoid a long or apologetic introduction, because it will dampen the students' interest in the lesson.

2. Development

This is the main part of the lesson. Here, the instructor develops the subject matter in a manner that helps the students achieve the desired learning outcomes.

The instructor must logically organize the material to show the relationships of the main points. The instructor usually shows these primary relationships by developing the main points in one of the following ways: from the past to the present, from the simple to the complex, from the known to the unknown, and from the most frequently used to the least frequently used.

a. *From Past to Present.* In this pattern of development, the subject matter is arranged chronologically, from the present to the past or from the past to the present. Such time relationships are most suitable when history is an important consideration, as in tracing the development of radio navigation systems.

b. *From Simple to Complex.* This pattern helps the instructor lead the student from simple facts or ideas to an understanding of involved phenomena or concepts. In studying jet propulsion, for example, the student might begin by considering the action involved in releasing air from a toy balloon and finish by taking part in a discussion of a complex gas turbine.

c. *From Known to Unknown.* By using something the student already knows as the point of departure, the instructor can lead into new ideas and concepts. For example, in developing a lesson on heading indicators, the instructor could begin with a discussion of the vacuum-driven directional gyro before proceeding to a description of the radio magnetic indicator (RMI).

d. *From Most Frequently Used to Least Frequently Used.* In some subjects, certain information or concepts are common to all who use the material. This fourth organizational pattern starts with common usages before progressing to the rarer ones. In learning navigation, students should study pilotage, dead reckoning, and VOR procedures before going on to more complex systems.

Under each main point in a lesson, the subordinate points should lead naturally from one to the other. With this arrangement, each point leads logically into, and serves as a reminder of, the next. Meaningful transitions from one main point to another keep the students oriented, aware of where they have been and where they are going.

Organizing a lesson so that the students will grasp the logical relationships of ideas is not an easy task. But this type of organization is necessary if the students are to learn. Poorly organized information is of little or no value to the student.

3. Conclusion

An effective conclusion retraces the important elements of the lesson and relates them to the objective. This review and wrap-up of ideas reinforces the student's learning and improves the retention of what has been learned. No new ideas should be introduced in the conclusion, because at this point they are likely to confuse the students.

LECTURE METHOD

Every instructor should know how to prepare and present a lecture and should understand the advantages and limitations of this teaching method. The lecture is used primarily to introduce students to a new subject, but it is also a valuable method for summarizing ideas, showing relationships between theory and practice, and reemphasizing main points. The lecture method is adaptable and has several advantages. Lectures may be given to either small or large groups; lectures may be used to introduce a complete training program or a unit of instruction; lectures may be combined with other teaching methods to give added meaning and direction.

Types of Lectures

Oral presentations may take several forms and may have various purposes. Among the most common forms are: (1) the illustrated talk in which the speaker relies heavily on visual aids to convey his ideas to the listeners; (2) the briefing in which the speaker presents a concise array of facts to the listeners who do not expect elaboration or supporting material; (3) the formal speech in which the speaker's purpose is to inform, to persuade, or to entertain; and (4) the teaching lecture for which the instructor must plan and deliver an oral presentation in a manner that helps the students reach the desired learning outcomes.

Teaching Lecture

The success of the teaching lecture depends upon the instructor's ability to communicate effectively with the class as well as the ability to plan, develop, and support the lesson. The instructor must determine the method of development to be used; i.e., *past to present*, *simple to complex*, etc. The instructor must also determine the depth of the ideas presented.

In other methods of teaching (demonstration-performance, guided discussion, etc.) the teacher receives direct reaction from the students in the form of verbal or motor activity; however, in the teaching lecture the feedback is not as direct and therefore much harder to interpret. In the teaching lecture, the instructor must develop a keen perception for subtle responses from the class (facial expressions, manner of taking notes, and apparent interest or disinterest in the lesson), and must be able to interpret the meaning of these reactions and adjust the lesson accordingly.

In developing a lesson, the instructor presents a number of main ideas or key points which support the overall objective and help the students visualize, know, or understand these points.

Preparing the Teaching Lecture

The competent instructor knows that careful preparation is one key to successful performance as a classroom lecturer. This preparation should start well in advance of the presentation. The following four steps should be followed in the planning phase of preparation: (1) establishing the objective and desired outcomes, (2) researching the subject, (3) organizing the material, and (4) planning productive classroom activities. In all stages of preparing for the teaching lecture, the instructor should support any point to be covered with meaningful examples, comparisons, statistics, or testimony.

In supporting key points or ideas in the lesson, the instructor must work on the assumption that the student may neither believe nor understand any point to be covered without the use of testimony from experts in the topic area or without meaningful examples, statistics, or comparisons to further explain the ideas. In developing the lesson, the instructor should also strongly consider the use of some personal ideas and concepts concerning the chosen subject of the lesson.

After completing the preliminary planning and writing of the lesson plan, the instructor should rehearse the lecture to build self-confidence. In rehearsals, or dry runs, the mechanics of using notes, visual aids, and other instructional devices can be smoothed out. If possible, the instructor should have another knowledgeable person, preferably another instructor, attend the practice sessions and observe the presentation critically. This critique will help the instructor judge the adequacy of supporting materials and visual aids.

Suitable Language

In the teaching lecture, simple rather than complex words should be used whenever possible. Good newspapers offer examples of the effective use of simple words. Picturesque slang and free-and-easy colloquialisms, if they suit the subject, can add variety and vividness to a teaching lecture. The instructor should not, however, use substandard English. Errors in grammar and vulgarisms detract from an instructor's dignity and reflect upon the intelligence of the students.

If the subject matter includes technical terms, the instructor should clearly define each one so that no student is in doubt about its meaning. Whenever possible, the instructor should use specific rather than general words. For example, the specific words "a leak in the fuel line" tell more than the general term "mechanical defect." Figurative language can add interest and color to a lecture.

Another way the instructor can enliven the lecture is to use sentences of varying lengths. The consistent use of short sentences results in a choppy style. But unless long sentences are carefully constructed, they are difficult to follow. Long sentences, inexpertly used, can become as tangled as a plate of spaghetti. To ensure clarity and variety, the instructor should use sentences of short and medium lengths.

Types of Delivery

The instructor can deliver a lecture in one of four ways: (1) by reading from a typed or written manuscript, (2) by reciting memorized material without the aid of a manuscript, (3) by speaking extemporaneously from an outline, or (4) by speaking impromptu without preparation.

The teaching lecture is probably best delivered in an extemporaneous manner. The instructor speaks from a mental or written outline but does not read or memorize the material to be presented. Because the exact words with which to express an idea are left to the moment, the lecture is more personalized than one which is read or spoken from memory. Since the instructor talks directly to the students, their reactions can be readily observed, and adjustments can be made to their responses. The instructor has better control of the situation, can change the approach to meet any emergency, and can tailor each idea to suit the responses of the students. For example, if the instructor realizes from their puzzled expressions that a number of students fail to grasp an idea, that point can be elaborated upon until the reactions of the students indicate that they understand.

The extemporaneous presentation reflects the instructor's personal enthusiasm and is more flexible than other methods. For these reasons, it is likely to hold the interest of the students.

Use of Notes

An instructor who is thoroughly prepared can usually speak effectively without notes. If the lecture and outline have been carefully prepared, and the instructor is completely familiar with both, there should be no real difficulty. However, the instructor whose preparation has been superficial, may find it necessary to use notes as a crutch.

AN INSTRUCTOR SHOULD NOT BE OVERLY DEPENDENT ON HIS LECTURE NOTES

Notes used widely can, however, have certain advantages. They assure accuracy, jog the memory, and dispel the fear of forgetting. They are essential for reporting complicated information. For a discursive, rambling instructor, notes are a must, because they help keep the lecture on the track. The instructor who requires notes should use them sparingly and unobtrusively, but at the same time should make no effort to hide them from the students. Notes should be written legibly or typed, and they should be placed on the lectern where they can be consulted easily, or held, if the instructor walks about the platform.

Formal Versus Informal Lectures

The lecture may be conducted in either a formal or an informal manner. The informal lecture includes active student participation. The primary consideration in the lecture method, as in all other teaching methods, is the achievement of desired learning outcomes. Learning is best achieved if students participate actively in a friendly, relaxed atmosphere. Therefore, the use of the informal lecture is encouraged. At the same time, it must be realized that a formal lecture is still to be preferred on some subjects and occasions, such as lectures introducing new subject matter.

The instructor can achieve active student participation in the informal lecture through the use of questions. In this way, the students are encouraged to make contributions that supplement the lecture. The instructor can use questions for one or more of the following purposes: to determine the experience and background of the students in order to tailor the lecture to their needs, to add variety and stimulate interest, and to check student understanding.

It is the instructor's responsibility to plan, organize, develop, and present the major portion of a lesson, and therefore he or she should not depend on the students for any significant portion of lesson development.

Advantages and Disadvantages of the Lecture

In a lecture, the instructor can present many ideas in a relatively short time. Facts and ideas that have been logically organized can be concisely presented in rapid sequence. Lecturing is unquestionably the most economical of all teaching methods in terms of the time required to present a given amount of material.

The lecture is particularly suitable for introducing a subject. To assure that all students have the necessary background to learn a subject, the instructor can present this basic information in a lecture. By using a lecture in this way, the instructor can offer students with varied backgrounds a common understanding of principles and facts.

The lecture is a convenient method for instructing large groups. If necessary, a public address system can be used to amplify the speaker's voice.

The lecture can be used to present information that would be difficult for the student to get in other ways. If the students do not have the time required for research or if they do not have access to reference material, the information they need can be presented to them by the lecture method.

The lecture can usefully and effectively supplement other teaching devices and methods. A brief introductory lecture can give direction and purpose to a demonstration. A lecture can also prepare students for a discussion by telling them something about the subject matter to be covered.

Although the lecture method can help the instructor meet the special challenges discussed above, it does have drawbacks. Too often the lecture does not provide for student participation and, as a consequence, many students willingly let the instructor do all the work. Learning is an active process, and the lecture method tends to foster passiveness and teacher-dependence on the part of the students. As a teaching method, the lecture does not bring about maximum attainment in certain types of learning outcomes. Motor skills, for example, can hardly be learned by listening to a lecture. The only way students can perfect such skills is through practice in performing them.

The lecture does not enable the instructor to estimate the students' progress before an explanation is given to them. Within a single period, the instructor may unwittingly present more information than students can absorb, and the lecture method provides no accurate means of checking student learning.

Many instructors find it difficult to hold the attention of all students in a lecture lasting throughout the class period. To achieve desired learning outcomes through the lecture method, an instructor needs considerable skill in speaking.

GUIDED DISCUSSION METHOD

In contrast to the lecture method, where the instructor provides information, the guided discussion method relies on the students to provide ideas, experiences, opinions, and information. An instructor may use this method during classroom periods, and preflight and postflight briefings, after the students have gained some knowledge and experience.

GUIDED DISCUSSION

Fundamentally, the guided discussion method is the reverse of the lecture method. The instructor aims to "draw-out" what the students know, rather than to spend the class period "telling them." The instructor must remember that the more intense the discussion and the greater the participation, the more effective the learning will be. The instructor must be sure that all members of the group follow the discussion, and that all are treated impartially. The instructor must encourage questions, exercise patience and tact, and comment on all responses. Sarcasm or ridicule should never be used.

Use of Questions in a Guided Discussion

In the guided discussion, learning is produced through the skillful use of questions. Questions can be categorized by function and by characteristics. Understanding these distinctions helps the instructor become a more skilled user of questions.

The instructor often uses a question to open up an area for discussion. This is the lead-off question and its function is indicated by its name. Its purpose is to get discussion started. After the discussion develops, the instructor may ask a follow-up question to guide the discussion. The reasons for using a follow-up question may vary. The instructor may want a student to explain something more thoroughly, or may need to bring the discussion back to a point from which it has strayed.

In terms of their characteristics, questions can be identified as *overhead*, *rhetorical*, *direct*, *reverse*, and *relay*. The overhead question is directed to the entire group to stimulate the thought and response from each group member. The instructor uses an overhead question to pose the lead-off question. The rhetorical question is similar in nature, because it also spurs group thought. However, the instructor answers the rhetorical question. Consequently, it is more commonly used in lecturing than in guided discussion.

The instructor who wants to phrase a question for follow-up purposes may choose the overhead type. If, however, a response is desired from a specific individual, a direct question may be asked of that student. The instructor may use a reverse question in response to a student's question. Rather than give a direct answer to the student's query, the instructor can redirect the question for the student to provide the answer. A relay question is redirected to the group instead of the individual.

A few guidelines for preparing effective questions follow. They should: (1) have a specific purpose, (2) be clear in meaning, (3) contain a single idea, (4) stimulate thought, (5) require definite answers, and (6) relate to previously taught information.

Planning a Guided Discussion

Planning a guided discussion is basically the same as planning a lecture. The instructor will find the following suggestions helpful in planning a discussion lesson.

Select a topic the students can profitably discuss. Unless the students have some knowledge to exchange with each other, they cannot reach the desired learning outcomes by the discussion method. If necessary, make assignments that will give the students an adequate background for discussing the lesson topic.

Establish a specific lesson objective and desired learning outcomes. Through discussion, the students develop an understanding of the subject by sharing knowledge, experiences, and backgrounds. Consequently, the objective normally is stated at the "understanding" level of learning. The desired learning outcomes should stem from, and be related to, the objective.

Conduct adequate research to become familiar with the topic. While researching, the instructor should always be alert for ideas on the best way to tailor a lesson for a particular group of students. Similarly, the instructor can prepare the pre-discussion assignment more effectively while conducting research for the classroom period. During this research process, the instructor should also earmark reading material that appears to be especially appropriate as student background material. Such material should be well organized and based on fundamentals.

Organize the main and subordinate points of the lesson in a logical sequence. The guided discussion has three main parts—introduction, discussion, and conclusion. The introduction consists of the attention step, motivation, and overview; the conclusion consists of the summary, remotivation, and closure. In the discussion, the instructor should be certain that the main points discussed build logically to the objective. By organizing in this manner, the instructor phrases the questions to help the students obtain a firm grasp of the subject matter and to minimize the possibility of a rambling discussion.

Plan at least one lead-off question for each desired learning outcome. In preparing questions, the instructor should remember that the purpose is to bring about discussion, not merely to get answers. The instructor should avoid questions that require only short categorical answers, such as "yes," "no," "green," "white," "one," or "four." Lead-off questions should usually begin with "how" or "why." It is better, for example, to ask, "Why does an airplane normally require a longer takeoff run at Denver than at New Orleans?" than to ask "Would you expect an airplane to require a longer takeoff run at Denver or at New Orleans?" Students can answer the second question by merely saying "Denver," but the first question is likely to start a discussion of air density, engine efficiency, the effect of temperature, etc.

Student Preparation for a Guided Discussion

It is the instructor's responsibility to help students prepare themselves for the discussion, and therefore he or she should encourage each student to accept responsibility for contributing to, and profiting from, the discussion.

Throughout the time the instructor prepares the students for their discussion, they should be made aware of the lesson objective. In certain instances, the instructor has no opportunity to assign preliminary work and must face the students "cold" for the first time. In such cases, it is practical and advisable to give the students a brief general survey of the topic during the introduction. Under no circumstances should students without some background in a subject be asked to discuss that subject.

Guiding a Discussion—Instructor Technique

Introduction. A guided discussion lesson is introduced in the same manner as the lecture. The introduction should include an attention step, a motivation step, and an overview of key points. To encourage enthusiasm and stimulate discussion, the instructor should create a relaxed, informal atmosphere. Each student should be given the opportunity to discuss the various aspects of the subject, and should be made to feel free to do so. Moreover, the student should feel a personal responsibility to contribute. The instructor can make the students feel that their ideas and active participation are wanted and needed.

Discussion. The instructor opens the discussion by asking one of the prepared lead-off questions. After asking a question, the instructor should be patient. The students should be given a chance to react. The instructor has the answer in mind before asking the question, but the student has to think about the question before answering. Thinking takes time. Sometimes an instructor finds it difficult to be patient while students figure out answers. It must be remembered that it takes time to recall data, word an answer, or to think of an example.

The more difficult the question, the more time the student will need to produce an answer. Sometimes students do not understand the question. Whenever the instructor sees puzzled expressions, the question should be stated again in a slightly different form. The nature of the questions should be determined by the lesson objective and desired learning outcomes.

Once the discussion is underway, the instructor should listen attentively to the ideas, experiences, and examples contributed by the students during the discussion. Remember that during the preparation, the instructor listed some of the anticipated responses that would, if discussed by the students, indicate that they had a firm grasp of the concept of the subject being discussed. As the discussion proceeds, the instructor may find it necessary to guide the direction, to stimulate the students to explore the subject in greater depth, or to encourage them to discuss the topic in more detail. By using "how" and "why" follow-up questions, the instructor should be able to guide the discussion toward the objective of the students to understand the subject.

When it appears the students have discussed the ideas that support this particular part of the lesson, the instructor should summarize what the students have accomplished. In a discussion lesson, the interim summary is one of the most effective tools available to the instructor. To bring ideas together and help in transition, an interim summary can be made immediately after the discussion of each learning outcome. This will summarize the ideas developed by the group and show how they relate to, and support, the idea discussed. The interim summary may be omitted after discussing the last learning outcome when it is more expedient for the instructor to present the first part of the conclusion. An interim summary reinforces learning in relation to a specific learning outcome. In addition to its uses as a summary and transitional device, the interim summary may also be used to keep the group on the subject or to divert the discussion to another member.

Throughout the discussion it is desirable to record ideas, facts, and agreements, so that the group can actually see relationships and the progress that has been made. A chalkboard and chalk or a large plain paper flip chart and grease pencil are suitable for this purpose.

Conclusion. A guided discussion is closed by summarizing the material covered. In the conclusion, the instructor should tie together the various points or topics discussed and show the relationships between the facts brought forth and the practical application of these facts. For example, in concluding a discussion on density altitude, an instructor might give a fairly complete description of an accident which occurred due to a pilot attempting to take off in an overloaded airplane from a small high-altitude airport on a hot day.

The summary should be brief but not to the point of incompleteness. If the discussion has revealed that certain areas are not understood by one or more members of the group, the instructor should clarify or cover this material again.

DEMONSTRATION-PERFORMANCE METHOD

This method of teaching is based on the simple, yet sound, principle that we learn by doing. Students learn physical or mental skills by actually performing those skills under supervision. An individual learns to write by writing, to weld by welding, and to pilot an airplane by actually performing flight maneuvers. Students also learn mental skills, such as speed reading, by this method. Skills requiring the use of tools, machines, and equipment are particularly well suited to this instructional method.

Every instructor should recognize the importance of student performance in the learning process. Early in a lesson that is to include demonstration and performance, the instructor should identify the most important learning outcomes; next, explain and demonstrate the steps involved in performing the skill being taught; and finally, allow the students time to practice each step, and thus increase their ability to perform the skill.

The demonstration-performance method is widely used. The science teacher uses it during laboratory periods, the aircraft maintenance instructor uses it in the shop, and the flight instructor uses it in teaching piloting skills. The demonstration-performance method of teaching has five essential phases: (1) explanation, (2) demonstration, (3) student performance, (4) instructor supervision, and (5) evaluation.

Explanation Phase

Explanations must be clear, pertinent to the objectives of the particular lesson to be presented, and based on the known experience and knowledge of the students. In teaching a skill, the instructor must convey to the students the precise actions they are to perform. In addition to the necessary steps, the instructor should describe the end result of these efforts. Before leaving this phase, the instructor should encourage the students to ask questions about any step of the procedure that they do not understand.

Demonstration Phase

The instructor must show the students the actions necessary to perform a skill. As little extraneous activity as possible should be included in the demonstration if the students are to clearly understand that the instructor is accurately performing the actions previously explained. If, due to some unanticipated circumstances the demonstration does not closely conform to the explanation, this discrepancy should be immediately acknowledged and explained.

Student Performance and Instructor Supervision Phases

Because these two phases, which involve separate actions, are performed concurrently, they are discussed here under a single heading. The first of these phases is the student's performance of the physical or mental skills that have been explained and demonstrated. The second activity is the instructor's supervision.

Student performance requires the students to act and do. To learn skills, students must practice. The instructor must, therefore, allot enough time for meaningful student activity. Through doing, the students learn to follow correct procedures and to reach established standards. It is important that students be given an opportunity to perform the skill as soon as possible after a demonstration. In flight training, the instructor may allow the student to "follow-through" on the controls during the demonstration of a maneuver. Immediately thereafter, the instructor should have the student attempt to perform the maneuver, coaching as necessary. If, during a classroom period, students have been performing an operation in unison (a weight and balance computation, for example), prior to terminating the performance phase they should be allowed to complete the operation at least once independently, with supervision and coaching on an "as needed" basis.

Evaluation Phase

In this phase, the instructor judges student performance. The student displays whatever competence has been attained, and the instructor discovers just how well the skill has been learned. To test each student's ability to perform, the instructor requires the students to work independently throughout this phase and makes some comment as to how each performed the skill relative to the way it was taught. From this measurement of student achievement, the instructor determines the effectiveness of the instruction.

PROGRAMMED INSTRUCTION

Interest in, and emphasis on, programmed instruction is constantly increasing among educators. This educational method is perceived by many authorities as the first major advance in education and training in years. Programming techniques are not altogether new to educational psychologists, but only within recent years have these been used in developing self-instructional materials in textbook form or for display in teaching machines.

As students progress through validated programmed instructional materials, they make a response to each increment of instruction. The material offers them an immediate feedback (knowledge of how well they are doing) by informing them of the correctness of their responses. The successful completion of each of these increments takes the students one step closer to the planned learning objectives. The major characteristics of programmed instruction are:

1. Clear specification of what the students must be able to do after training.
2. Careful sequencing of material.
3. Presentation of material in steps which challenge students but do not exceed their ability.
4. Active student responses.
5. Immediate confirmation of answers.

6. Test and revision of instructional material until it has been proved able to teach to the desired objective level.

This approach systematically carries students, step by step, to the learning objectives they are to attain. In this respect, programmed instruction is generally more tutorial in nature than the instruction in a typical classroom situation. That is, programmed instruction not only gives students what they are to learn but also guides them in the way they are to learn.

One Method of Programming

A number of methods have been used in developing programmed instructional materials. The linear method described here was developed by psychologist B. Frederick Skinner and is based largely on reinforcement (rewarding the student) for accurate performance. According to Skinner, in ideal programmed instruction the student cannot make an error. In linear-programmed instruction, the material is itemized and presented in small steps. A student is prompted, or cued, as necessary, so that almost invariably the correct response is given. Materials are carefully designed to offer as much review as needed to assure the degree of retention appropriate to the subject matter, the learning situation, and the needs of the students.

The student responds to a frame (the printed matter for one step or increment in the instruction) usually by writing words into spaces provided for that purpose. Linear-programmed material may, however, be designed to elicit other types of responses. Answers may be given mentally or orally. Simple tasks may be performed. Sequences of more complicated acts that make up a complete procedure may be required.

After completing the response to a frame, the student immediately confirms the correctness of the response by comparing it to the program answer before continuing on to the next frame. Thus, the student progresses smoothly with a continuous awareness of being correct giving a sense of satisfaction. If the programmer has met the challenge of properly constructing materials in a linear format, the student will, at a comfortable rate and almost effortlessly, learn the material presented.

Proponents of this system attribute its success to the reinforcement it provides and the repetition it uses. If a student encounters the same fact, idea, or concept in a number of ways and if reinforcement or reward occurs each time a correct answer is made, learning takes place. Apparently, the more often the student is rewarded, the more lasting the learning.

In each block of new subject matter, initial frames contain obvious cues to correct responses. Thus, a student finds it virtually impossible to make errors. In subsequent frames, cues are gradually withdrawn and finally eliminated. As a student approaches each terminal behavior, or learning objective, cues are withdrawn until the student supplies complete answers without being cued.

The following sequence of frames illustrates the linear method. These sample frames are typical of initial frames. They offer obvious cues and require only simple responses.

To the casual observer, this sequence may seem unduly simple. To the student who is totally unfamiliar with the subject matter, however, it offers a sort of "learning game."

1. In straight-and-level unaccelerated flight, an airplane is acted on by four forces—lift, weight, thrust, and drag. In this stabilized flight condition weight equals ________.
1. lift. 2. In straight-and-level unaccelerated flight, thrust equals ________.
2. drag. 3. An increase in the velocity of air passing over a wing results in an increase of both ________ and drag.
3. lift. 4. When lift exceeds weight, an airplane will enter a ________.
4. climb. 5. . . . . .

CHAPTER VI. THE INSTRUCTOR AS A CRITIC

Although this chapter deals with the critique primarily from the standpoint of the instructor in the classroom, the techniques and methods described also apply to the aircraft maintenance instructor in the shop and to the flight instructor/student pilot situation.

No skill is more important to an instructor than the ability to analyze, appraise, and judge student performance. The student quite naturally looks to the instructor for guidance, analysis, appraisal, suggestions for improvement, and encouragement.

A critique may be either oral, or written, or both. It should come immediately after a student's individual or group performance, while the details of the performance are easy to recall. An instructor may critique any activity which a student performs or practices to improve skill, proficiency, and learning. A critique may be conducted in private or before the entire class. A critique presented before the entire class can be beneficial to every student in the classroom as well as to the student who performed the exercise or assignment.

Two common misconceptions about the critique should be corrected at the outset. *First*, a critique is not a step in the grading process. It is a step in the learning process. *Second*, a critique is not necessarily negative in content. It considers the good along with the bad, the whole in terms of its parts, or the parts in relation to the whole, or to each other. A critique can be as many-sided and varied in content as the performance.

PURPOSE OF A CRITIQUE

A critique should improve the students' performance and provide them with something constructive with which to work and on which they can build. It should provide direction and guidance to raise their level of performance. Students must understand the purpose of the critique; otherwise, they cannot accept the criticism offered and little improvement can be expected.

A critique can be used as a re-teaching device. Not all critiques lend themselves to re-teaching, but the instructor should be alert to the possibilities and take advantage of such an opportunity when it arises. If, for example, several students falter when they reach the same step in a weight and balance problem, the instructor might recognize a need for re-explaining, or re-demonstrating the step, or giving it special emphasis in the critiques of subsequent performance.

CHARACTERISTICS OF AN EFFECTIVE CRITIQUE

A Critique Should Be Objective

The effective critique is focused on student performance and should not reflect the personal opinions, likes, dislikes, and biases of the instructor. For example, if a student accomplishes a complicated flight planning problem, it would hardly be fair for the instructor to criticize the student's personality traits unless they interfered with the performance itself. Critiquers sometimes permit their judgments to be influenced by their general impressions of the student, favorable or unfavorable. Sympathy for, or overidentification with, a student to such a degree that it influences objectivity is known as "error of halo." A conflict of personalities can also color an opinion. If a critique is to be objective, it must be honest; it must be based on the performance as it was, not as it could have been, or as the instructor and student wish that it had been.

A Critique Should Be Flexible

Obviously, a critique cannot be made until the performance is finished. Sometimes a good student will turn in a poor performance and a poor student will turn in a good one. A friendly student may suddenly become hostile, or a hostile student may suddenly become friendly and cooperative. The instructor must fit the tone, technique, and content of the critique to the occasion and the student. A critique should never be designed and executed so that the instructor cannot allow for variables. Again and again, the instructor is faced with the problem of what to say, what to omit, what to stress, what to minimize—and it is the peculiar challenge of the critique that causes the instructor to determine what to say at the proper moment. An effective critique is one that is flexible enough to satisfy the requirements of the moment.

A Critique Should Be Acceptable

Before students willingly accept their instructor's criticism, they must first accept the instructor. The students must have confidence in the instructor's qualifications, teaching ability, sincerity, competence, and authority. Usually, instructors have the opportunity to establish themselves with their students before the formal critiquing situation arises. If this is not the case, however, the instructor's manner, attitude, and readily apparent familiarity with the subject at hand must serve instead. Critiques do not have to be all sweetness and light, nor do they have to curry favor with students. If a critique is fair, is presented with authority, conviction, sincerity, and from a position of recognizable competence, the student probably accepts it as such. Wise instructors do not rely on their position to make a critique more acceptable to their students. While such factors usually operate to the instructor's advantage, acceptability depends on more active and demonstrable qualities.

A Critique Should Be Comprehensive

A comprehensive critique is not necessarily a long one, nor must it treat every aspect of the performance in detail. The instructor must decide whether the greater benefit will come from a discussion of a few major points or a number of minor points. The criteria may be what most needs improvement, or only what the student can reasonably be expected to improve. An effective critique covers strengths as well as weaknesses. How to balance the two is a decision that only the instructor can make. To dwell on the excellence of a performance to the neglect of that portion that should be improved is a disservice to the student.

A Critique Should Be Constructive

A critique is pointless unless the student profits from it. Praise for praise's sake is of no value if a student is not taught how to capitalize on things which are done well and to use them to compensate for lesser accomplishments. By the same token, it is not enough to identify a fault or weakness. The instructor should give positive guidance for correcting the fault and strengthening the weakness. Negative criticism that does not point toward improvement or a higher level of performance could be omitted from a critique altogether.

A Critique Should Be Well Organized

Unless a critique follows some pattern of organization, a series of otherwise valid comments may lose their impact. Almost any pattern is acceptable so long as it is logical and makes sense to the student as well as to the instructor. An effective organizational pattern might be the sequence of the performance itself. Sometimes a critique can profitably begin with the point where a demonstration failed and work backward through the steps that led to the failure. A success can be analyzed in similar fashion. Sometimes a defect is so glaring or a strength so great that it overshadows the rest of the performance and can serve as the core of a critique. Breaking the whole into parts or building the parts into a whole has strong possibilities. Whatever the organization of the critique, the instructor should be flexible enough to change it if the student cannot follow and understand it.

A Critique Should Be Thoughtful

An effective critique reflects the instructor's thoughtfulness toward the student's need for self-esteem, recognition, and approval from others. The critiquer should never minimize the inherent dignity and importance of the individual. Ridicule, anger, or fun at the expense of the student have no place in a critique. On occasion, an instructor should criticize the student in private. In some cases, discretion may rule out any criticism at all. For example, criticism does not help a student whose performance is impaired by a physiological defect. While being straightforward and honest, the instructor should always respect the student's personal feelings.

A Critique Should Be Specific

The instructor's comments and recommendations should be specific, not so general that the student can find nothing to hold on to. A statement such as, "Your second weld wasn't as good as your first," has little constructive value unless the student learns specifically wherein it was not so good. If the instructor has a clear, well-founded, and supportable idea in mind, it should be expressed with firmness and authority in terms that cannot be misunderstood. Students cannot act on recommendations unless they know specifically what the recommendations are. At the conclusion of a critique, students should have no doubt what they did well and what they did poorly and, most importantly, specifically how they can improve.

METHODS OF CRITIQUE

The critique of student performance is always the instructor's responsibility, and it can never be delegated in its entirety. The instructor can add interest and variety to the criticism, however, through the use of imagination and by drawing on the talents, ideas, and opinions of others. The following are useful methods of conducting a critique.

Instructor-Student Critique

The instructor leads a group discussion in which members of the class are invited to offer criticism of a performance. This method should be controlled carefully and directed with firm purpose. It should be organized and not allowed to degenerate into a random, free-for-all.

Student-Led Critique

The instructor asks a student to lead the critique. The instructor can specify the pattern of organization and the techniques or can leave it to the discretion of the student leader. Because of the inexperience of the participants, student-led critiques may not be efficient, but they can generate student interest and learning and, on the whole, be effective.

Small-Group Critiques

The class is divided into small groups and each group is assigned a specific area to criticize. These groups must present their findings to the class. Frequently, it is desirable for the instructor to furnish the criteria and guidelines. The combined reports from the groups can make for a comprehensive critique.

Individual Student Critique

The instructor requires a student to present the entire critique. A variation is for the instructor to put questions to a number of students (socialized recitation) about the manner and quality of performance.

Written Critique

Written critiques have three advantages. *First*, the instructor can devote more time and thought to it than to an oral critique in the classroom. *Second*, the students can keep written critiques and refer to them whenever they wish. *Third*, when the instructor requires all the students to write a critique of a performance, the student-performer has the permanent record of the suggestions, recommendations, and opinions of all the other students. The disadvantage is that other members of the class do not benefit from the criticism.

Self-Critique

A student is required to criticize personal performance. Like all other methods, a self-critique must be controlled and supervised by the instructor. Whatever the methods employed, the instructor must not leave controversial issues unresolved nor erroneous impressions uncorrected. If the students participate in the critique, the instructor must make allowances for their relative inexperience. Normally, the instructor should reserve time at the end of the student critique to cover those areas that might have been omitted or not emphasized sufficiently, or areas that are considered worth repeating.

GROUND RULES FOR CRITIQUING

1. Except in rare and unusual instances, do not extend the critique beyond its scheduled time and into the time allotted for other activities. A point of diminishing returns can be reached quickly.
2. Avoid trying to cover too much. A few well-made points may be more beneficial than a large number of points that are not developed adequately.
3. Allow time for a summary of the critique itself to reemphasize the most important things a student should remember.
4. Avoid dogmatic or absolute statements, remembering that most rules have exceptions. A critiquer can be wrong.
5. Avoid controversies with the class, and do not get into the delicate position of taking sides with group factions.
6. Never allow yourself to be maneuvered into the unpleasant position of defending criticism. If the criticism is honest, objective, constructive, and comprehensive, no defense should be necessary.
7. If part of the critique is written, make certain that it is consistent with the oral portion.

CHAPTER VII. EVALUATION

Evaluation is an integral part of the learning process. Whenever learning takes place, the result is a definable, observable, measurable change in behavior. Evaluation is concerned with defining, observing, and measuring or judging this new behavior. True evaluation is a continuous process. Once instruction has begun, some sort of evaluation is essential to determine both *what* and *how well* the students are learning. The instructor's evaluation may consist of simple observations of the students' performance or it may be accomplished by oral quizzing, by administering written tests, or by performance tests.

ORAL QUIZZING

The most practical means of evaluation is the direct or indirect oral questioning of students by the instructor. Questions may be loosely classified as fact questions and thought questions. The answer to a fact question is based on memory or recall. This type of question usually concerns *who*, *what*, *when*, or *where*. Thought questions usually involve *why* or *how*, and require the student to combine a knowledge of facts with an ability to analyze situations, solve problems, and arrive at conclusions.

Proper quizzing by the instructor can have a number of desirable results:

1. It reveals the effectiveness of the instructor's training procedures.
2. It checks the student's retention of what has been learned.
3. It reviews material already covered by the student.
4. It can be used to retain the student's interest and stimulate thinking.
5. It emphasizes the important points of training.
6. It identifies points which need more emphasis.
7. It checks the student's comprehension of what has been learned.
8. It promotes active student participation, which is important to effective teaching.

Characteristics of Effective Questions

Effective oral quizzing requires preparation; therefore, the instructor should devise and write pertinent questions in advance. One method is to place them in the lesson plan. These prepared questions serve merely as a framework and, as the lesson progresses, should be supplemented by such impromptu questions as the instructor considers appropriate. Usually an effective question has only *one correct answer.* This is always true of good questions of the objective type and generally true of all good questions, although the one correct answer to a thought question may be expressed sometimes in a variety of ways. To be effective, questions *must apply* to the subject of instruction. Unless the question pertains strictly to the particular training being conducted, it serves only to confuse the students and divert their thoughts to an unrelated subject. An effective question should be *brief* and *concise*, but must be clear and definite. Enough words must be used to establish the conditions or situations exactly, so that instructor and students will have the same mental picture.

To be effective, questions must be adapted to the ability, experience, and stage of training of the students.

Effective questions *center on only one idea.* One idea—one question. A single question should be limited to who, or what, or when, or where, or how, or why—not a combination.

Effective questions must present a *challenge* to the students. To be good, a question must be difficult—that it, difficult for the students at that particular stage of training. Questions of suitable difficulty serve to stimulate learning. Effective questions demand and deserve the use of *good English.*

Types of Questions to Avoid

Asking "Do you understand?", or "Have you any questions?", has no place in effective quizzing. Assurance by the students that they do understand or that they have no questions provides no evidence of their comprehension, or that they even know the subject which is under discussion.

Other typical types of questions which must be avoided are:

The puzzle—"What is the first action you should take if a conventional gear airplane with a weak right brake is swerving left in a right crosswind during a full-flap power-on wheel landing?"

The oversize—"What do you do before beginning an engine overhaul?"

The toss-up—"In an emergency, should an air crewmember activate the escape slide or control the crowd?"

Bewilderment—"In reading the altimeter—you know you set a sensitive altimeter for the nearest station pressure—if you take temperature into account, as when flying from a cold air mass through a warm front, what precaution should you take when in a mountainous area?"

Catch questions should be avoided at all times. The students will soon develop the feeling that they are engaged in a battle of wits with the instructor, and the whole significance of the subject of the instruction involved will be lost.

Irrelevant questions should be avoided. The teaching process must be an orderly procedure of building one block of learning upon another in orderly progression, until a desired goal is reached. Diversions, and the introduction of unrelated facts and thoughts, will only obscure this orderly process and retard the student's progress. Answers to unrelated questions are not helpful in evaluating the student's knowledge of the subject at hand.

Answering Students' Questions

The answering of students' questions must also conform with certain considerations if it is to be an effective teaching method. The question must be clearly understood by the instructor before an answer is attempted. The instructor should display interest in the student's question and frame as direct and accurate an answer as possible. After the instructor completes a response, it should be determined whether or not the student's request for information has been completely answered, and if the student is satisfied with the answer.

Sometimes it may be unwise to introduce the more complicated or advanced considerations, necessary to completely answer a student's question, at the current point in training. In this case, the instructor should carefully explain to the student that the question was good and pertinent but that the answer would, at this time, unnecessarily complicate the learning tasks at hand. The instructor should advise the student to reintroduce the question later at the appropriate point in training, if it has not in the meantime been resolved in the normal course of instruction.

Occasionally, a student asks a question which the instructor cannot answer. In such cases, the instructor should freely admit not knowing the answer, but should promise to get the answer or, if practicable, offer to help the student look it up in available references.

EFFECTIVE QUESTIONING IS A TWO-WAY PROCESS

In all quizzing conducted as a portion of the instruction process, "yes" and "no" answers should be avoided. Questions must be framed so that the answers may be specific and factual, but one-word answers may well be the product of a good guess and not truly representative of the learning or ability of the student. This applies to instructors' answers to students' questions, and to students' answers to quizzes used in the furtherance of training.

WRITTEN TESTS

As evaluation devices, written tests are only as good as the knowledge and proficiency of the test writer. In the following pages, an attempt has been made to provide the aviation instructor with some of the basic concepts of written test design. Since many excellent publications are available to the aviation instructor on test administration, test scoring, grade assignment, whole test analysis, and test item analysis, these topics are not covered in this chapter.

Characteristics of a Good Test

If a test is to be effective, it must have *reliability*, *validity*, *usability*, *comprehensiveness*, and *discrimination.*

Reliability. A reliable measuring instrument, including a written test, is one which yields consistent results. If identical measurements are obtained every time a certain instrument is applied to a certain dimension, the instrument is considered reliable. An unreliable instrument cannot be depended upon to yield consistent results. An altimeter that has worn moving parts, a steel tape that expands and contracts with temperature changes, or cloth tapes that are affected by humidity cannot be expected to yield reliable measurements.

No instrument is perfectly reliable. But it is obvious that some instruments are more reliable than others; for example, laboratory balances are more reliable than those used in grocery stores.

The reliability of an instrument can be estimated by taking many measures of the same item. For example, a rough measure of the reliability of a thermometer can be obtained by taking several readings of the temperature of a fluid held at a constant temperature. Except for the errors made by the person taking the readings, the amount of scatter in the readings can be considered as the degree of unreliability in the thermometer.

Reliability has the same meaning whether applied to written tests or to balances, thermometers, and altimeters. The term refers only to the consistency of the results obtained from the instrument.

Validity. A measuring instrument, including a written test, is valid when it actually measures what it is supposed to measure and nothing else.

The fact that an instrument is highly reliable does not necessarily mean that it is valid. The instrument which has maximum consistency (high reliability) may not be measuring what it is intended to measure. That is to say, an instrument may have very high reliability and low validity at the same time. For example, the butcher who hides a lump of solder under the pan of the scales might get reliable measurements but invalid results (at least from the customer's point of view).

If a mechanic wishes to measure the diameter of a bearing with a micrometer, it must be determined that the contacting surfaces of the bearing and the micrometer are free of grease and dirt. Otherwise, the measurement will be of both the diameter of the bearing and the extraneous matter. The measurements will be invalid.

A test used in educational evaluation follows the same principles of validity. If a test is intended to measure the ability of students to apply knowledge, it must measure application, and not the ability to recall and write down facts.

Evaluations used in the classroom are valid only to the extent that they measure achievement of the objectives of instruction.

A rough estimate of the validity of a classroom test may be obtained from the judgment of several competent instructors. To estimate validity, they read the test critically and consider its content in relation to the stated objectives of the instruction. Items which do not pertain directly to the objectives of the course are modified or eliminated.

Because it is related to purpose, validity is the most important feature of any written test. The instructor must carefully consider whether the test actually measures what it is supposed to measure.

Usability. A measuring instrument, including a written test, should be usable. A usable written test is easy to give; it is written in a type size large enough for the students to read easily; the wording of both the directions for taking the test and of the test items themselves is clear and concise; graphs, charts, and illustrations are appropriate to the test items and are clearly drawn; and the test is easily graded.

Comprehensiveness. A measuring instrument, including a written test, must sample liberally whatever is being measured. Suppose, for example, the owner of a grain elevator wants to evaluate a carload of wheat. If a bushel of wheat from the top of the carload is analyzed, the test would be poor—one that is not sufficiently comprehensive. A judgment based on this evaluation could not be safely applied to the entire carload; the wheat at the bottom of the car might be infested with insects. However, if 50 test tubes of wheat are taken from scattered positions throughout the carload, less than a bushel of wheat might be analyzed, but a much more comprehensive evaluation of the contents of the car would be obtained.

In classroom evaluation, a test must sample liberally the objectives of instruction. The comprehensiveness of a test is the degree to which the *scope* of the course objectives is tested. At best, however, any evaluation is but a sample of the entire course. Just as the owner of the wheat has to select samples of wheat from scattered positions in the car, the instructor has to make certain that the evaluation includes a representative and comprehensive sampling of the objectives of the course. In both instances, the evaluators must deliberately take comprehensive samples.

Discrimination. A measuring instrument, including a written test, must be constructed in such a way that it will detect small differences. Suppose, for example, that a machinist wishes to measure six bearings that are slightly graduated in size. If a ruler is used to measure the diameters of the bearings, little difference will be found between the smallest bearing and the second smallest one. If the machinist compares the third bearing with the first bearing, slight differences in size might be detected, but the ruler could not be depended on for accurately assorting the six bearings. However, if the machinist measures with a micrometer, which can measure very fine graduations, the diameters of the first and second bearing, the second and third bearing, and so on, can be easily differentiated.

In classroom evaluation, a test must be able to measure small differences in achievement in relation to the objectives of the course. When a test is constructed to identify the difference in the achievement of students, it has three features: (1) there is a wide range of scores; (2) all levels of difficulty are included; (3) each item distinguishes between the students who are low and those who are high in achievement of the course objectives.

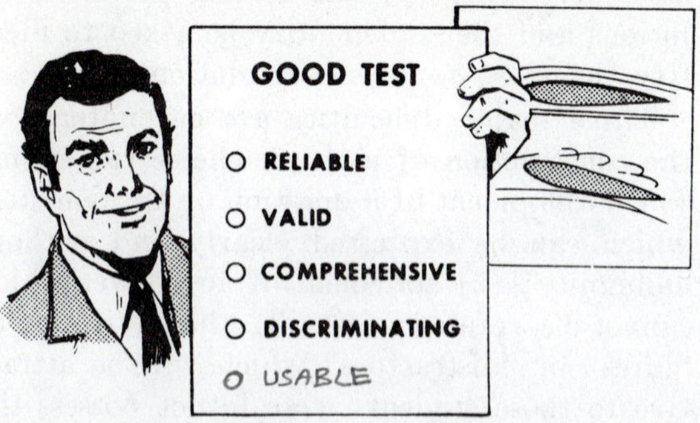

Written Test Items

Test items fall into two general categories, the *supply-type item* and the *selection-type item.* Items of the first category require the student to furnish a response in the form of a word, sentence, or paragraph. The second category includes items for which two or more alternative responses are provided.

Supply-type Test Items. This type item may be required where no selection-type items can be devised to secure adequate evaluation of student accomplishment. The supply-type item requires the students to organize their knowledge. It demands an ability to express ideas that is not required for a selection-type item. It permits the students to tell all they know about a subject even though they do not know the complete answer. This type item is valuable in measuring the students' generalized understanding of a subject.

On the other hand, a supply-type item may evaluate the students' ability to write rather than their specific knowledge of subject matter. It places a premium on neatness and penmanship. The main disadvantage of tests containing supply-type items is that they cannot be graded with uniformity. There is no assurance that the grade assigned is the grade deserved by the student. The same test graded by different instructors would probably be assigned different scores. The same test graded by the same instructor on consecutive days might be assigned two altogether different scores. Still another disadvantage of a test consisting of supply-type items is the time required by the student to complete it and the time required by the instructor to grade it. Everything considered, the disadvantages of the supply-type item appear to exceed its advantages to such an extent that most instructors prefer to use the selection-type item whenever possible.

Selection-type Test Items. Written tests made up of selection-type items are highly objective. That is, the results of such a test would be graded the same regardless of the student taking the test or the person grading it. Tests which include only selection-type items make it possible to directly compare student accomplishment. For example, it is possible to compare the performance of students within the same class or in different classes, students under one instructor with those under another instructor, student accomplishment at one stage of instruction with that at a later stage, or to compare student learning under one method of instruction to that under another method. By using selection-type items, the instructor can test on many more areas of knowledge in a given time than could be done by requiring the student to supply written responses. This increase in comprehensiveness can be expected to increase validity and discrimination. Another advantage is that selection-type test items are well adapted to statistical item analysis.

"True-False" Test Items. This item-type, with all its variations, has a wide range of usage. It is well adapted to the testing of knowledge of facts and details, especially when there are only two possible answers. The chief disadvantage is that the probability of guessing, in the simple true-false item, is greater than in any other type test item.

True-false test items are probably used and misused more than any other selection-type item. Frequently, instructors merely score textual material, select sentences more or less at random, and make half of them false by inserting negatives. They mistakenly believe they can justify these answers by referring students to pages in the text.

When tests are constructed in this way, the principal attribute being measured is photographic memory. Such test construction has aroused antagonism toward selection tests in general and true-false items in particular. It has also decreased the validity of educational evaluations.

Some of the principles which should be followed in the construction of true-false items are listed below:

1. Include only one idea in each statement.
2. Avoid the unnecessary use of negatives. They tend to confuse the reader.
3. Avoid involved statements. Keep wording and sentence structure as simple as possible. Make statements both definite and clear.
4. Whenever possible, use terms which mean the same thing to all students.
5. Avoid absolutes such as "all," "every," "only," "no," and "never." Since unequivocally true or false statements are rare, statements containing absolutes are usually false. Similarly, avoid statements containing "some," "any," and "generally." These words are known as *determiners* and provide clues to the correct answer.
6. Avoid patterns in the sequence of correct responses, because students can often identify the patterns. Instructors sometimes deliberately use patterns to make hand scoring easier. This is a poor practice.
7. Make statements brief and about the same length. Some instructors unconsciously make true statements longer than false ones. Students are quick to take advantage of this tendency.

"Multiple-Choice" Test Items. When properly devised and constructed, this item-type offers several unique advantages which make it more widely used and versatile than either the matching or the true-false item.

This type of item may be used to determine student achievement, ranging from acquisition of facts to understanding, reasoning, and ability to apply what has been learned. It is appropriate to use when the problem: (1) has a built-in and unique solution; e.g., specific application of laws or principles; (2) may be clearly limited by the wording of the item, so that the student must choose the *best* of several offered solutions rather than a universal solution; (3) is such that several options are plausible, or even scientifically accurate, but the student may be asked to identify the one *most pertinent;* (4) has several pertinent solutions and the student may be asked to identify the *most appropriate* solution.

Three major difficulties are encountered in the construction of multiple-choice test items: (1) development of a question or an item stem which can be expressed clearly and without ambiguity; (2) statement of an answer which cannot be refuted; and (3) the invention of "lures" or "distractors" which will be attractive to those students who do not possess the knowledge or understanding necessary to recognize the correct answer.

A multiple-choice item may take several basic forms: (1) it may be a direct question followed by several possible answers; (2) it may be an incomplete sentence followed by several possible completions of that sentence; or (3) it may be a stated problem, a graph, a diagram, etc., followed by several correct or incorrect statements covering that problem, graph, or diagram. The student may be asked to select the one choice which is the *correct* answer or completion, the one choice which is an *incorrect* answer or completion, or the *one* choice which is the *best of the answers presented in the test item.* These three methods of answering, combined with the three item forms, give the test constructor nine possible ways of setting up a multiple-choice question to measure a given ability. Experience has shown, however, that beginners are more successful with the question form.

Examples of Multiple-Choice Item Forms: Stem Presented as a Question. This form is generally better than the incomplete stem in that it is simpler and more natural.

What gas forms the largest part of the atmosphere?

1. Oxygen.
2. Nitrogen.
3. Helium.
4. Hydrogen.

Stem as an Incomplete Statement. When using this form, care must be exercised to avoid ambiguity, giving clues, and using unnecessarily complex or unrelated alternatives.

The atmosphere is a mixture of gases, the largest part being

1. oxygen.
2. nitrogen.
3. helium.
4. hydrogen.

Multiple Response Required. Students are instructed to select all correct answers.

What two gases make up the largest part of the atmosphere?

A. Oxygen
B. Nitrogen
C. Helium
D. Hydrogen
E. Neon
1—A and B
2—B and D
3—C and D
4—A and E

Stem Supplemented by an Illustration. Useful for measuring ability to read instruments, identify objects, etc.

What is the indicated altitude?

1. 440 feet
2. 4,000 feet
3. 4,400 feet
4. 40,000 feet

"None of the Above" or "All of the Above" as Alternatives. No sample is given for this form. It may be used when the correct alternative seems to give itself away in a set of ordinary alternatives. To be effective, these alternatives must sometimes be the correct response.

The Negative Variety. Always emphasize the negative word.

Which of the following is *not* used to control a helicopter in flight?

1. Cyclic
2. Collective pitch
3. Anti-torque pedals
4. Aileron

Association Type. This is useful if a limited number of associations are to be made. Matching items serve better if a large number of related associations are to be made.

Which maneuver does *not* belong with the other three?

1. Chandelle
2. Autorotation
3. Lazy eight
4. Pylon eight

Definition Type.

The erratic movement of the magnetic compass card due to turbulence or rough control technique is known as

1. acceleration error.
2. deceleration error.
3. oscillation error.
4. northerly turning error.

In general, the form with the alternatives as answers to a question is preferable to the form which uses an incomplete statement as the stem. It is more easily phrased and is more natural for the student to read. Less likely to contain ambiguities, it usually results in more homogeneity among the alternatives and gives fewer clues to the correct response.

When multiple-choice items are used, four or five alternatives are generally provided. It is usually difficult to construct more than five plausible responses (responses which appear to be correct to one who has not mastered the subject matter). If there are less than four alternatives, the probability of guessing the correct response is considerably increased.

Students are not supposed to guess the correct alternative; they should select it only if they know it is correct. It is, therefore, ethical to mislead the unsuccessful student into selecting an incorrect alternative. An effective and valid means of diverting the student from the correct response is to use common errors as distracting alternatives.

Items intended to measure the knowledge level of learning should have but one correct alternative; all other alternatives should be clearly incorrect. When items are to measure achievement at a higher level of learning, some or all of the alternatives should be acceptable responses—but one should be clearly better than the others. In either case, the instructions given should direct the student to select the best alternative.

Some of the principles which should be followed in the construction of multiple-choice items are listed below:

1. Make each item independent of every other item in the test. Do not permit any item to reveal the correct answer to another item or permit the solution to any item to depend upon knowing the correct solution to another item. If

items are allowed to be interdependent it becomes impossible to pinpoint specific deficiencies in either students or instruction.

2. Design items which call for essential knowledge rather than for abstract background knowledge or unimportant facts.
3. State each item in the working language of the student. Failure to do so can result in decreased validity of the test, since the ability to work with the language will be measured as well as the subject-matter knowledge or achievement.
4. Include sketches, diagrams, or pictures when they can present a situation more vividly than words. They generally speed the testing process, add interest, and help to avoid reading difficulties and technical language. A common criticism of written tests is the emphasis on the reading ability of the student. The validity of the examination is decreased when this ability is not an objective of the course or test.
5. When a negative is used, emphasize the negative word or phrase. A student who is pressed for time may identify the wrong response simply because the negative form is overlooked. To whatever extent this occurs, the validity of the test is decreased.
6. Items containing double negatives invariably cause confusion. If a word such as "not" or "false" appears in the stem, avoid using another negative word in the stem or in any of the alternatives.
7. Catch questions, unimportant details, ambiguities, and leading questions should be avoided, for they do not contribute to effective evaluation in any way. Instead, they tend to confuse and antagonize the student.

Stems. In preparing the stem of a multiple-choice item or in reviewing an old one, the following general principles should be kept in mind:

1. The stem of the item should clearly present the central problem or idea. The function of the stem is to set the stage for the alternative which follows.
2. The stem should contain only material relevant to its solution (unless the selection of what is relevant is part of the problem).
3. The stem should be worded in such a way that it does not give away the correct response. Avoid clue words or phrases (determiners).
4. Put everything that pertains to all alternatives in the stem of the item. This helps to avoid repetitious alternatives and saves time.
5. Generally avoid using "a" or "an" at the end of the stem. They may give away the correct choice. Every alternative should fit grammatically with the stem of the item.

Alternatives. In preparing or reviewing the alternatives to a multiple-choice item, the following principles are important to remember:

The alternatives in a multiple-choice test item are as important as the stem. They should be formulated with care; incorrectness should not be the only criterion for the distracting alternatives. Some distractors which can be used are (1) a response which is related to the situation and which sounds plausible to the untutored, but which is incorrect; (2) a common misconception; (3) a statement which itself is true but does not satisfy the requirements of the problem; or (4) a statement which is either too broad or too narrow for the requirements of the problem.

Keep all alternatives of approximately equal length. Research of instructor-made tests reveals that, in general, correct alternatives are longer than incorrect ones.

When alternatives are numbers, they should generally be listed in ascending or descending order of magnitude.

"Matching" Test Items. This type item is, in reality, a collection of related multiple-choice items. In a given period of time, more samples of a student's abilities can be measured using matching rather than by using equivalent multiple-choice items. The matching item is particularly good for measuring students' ability to recognize relationships and to make associations between terms, parts, words, phrases, clauses, or symbols listed in

one column with related items in another column. Matching reduces the probability of guessing correct responses—as compared to a series of multiple-choice items covering the same material—especially if alternatives may be used more than once. The testing time can also be used more efficiently. Samples of two different forms of this item follow.

Equal Columns. When using this form, always provide for some items in the response column to be used more than once, or not at all, to preclude guessing by elimination.

Directions: In the blank before each phrase in the left-hand column, write the letter corresponding to the abbreviation which is most closely associated with that phrase. Each abbreviation may be used more than once; some abbreviations may not be used at all.

—1. Never exceed speed	a. V_a
—2. Best angle-of-climb speed	b. V_{so}
—3. Red radial line on airspeed indicator	c. V_x
—4. Design maneuvering speed	d. V_{ne}
—5. Best rate-of-climb speed	e. V_y

Unequal Columns. Generally preferable to equal columns.

Directions: In the blank before each phrase in the left-hand column, write the letter corresponding to the abbreviation which is most closely associated with that phrase. Each abbreviation may be used more than once; some abbreviations may not be used at all.

—1. Never exceed speed	a. V_a
—2. Best angle-of-climb speed	b. V_{so}
—3. Red radial on airspeed indicator	c. V_x
	d. V_{ne}
—4. Design maneuvering speed	e. V_y
—5. Best rate-of-climb speed	f. V_{le}

Some of the principles which should be followed in the construction of matching items are listed below:

1. Give specific and complete instructions. Do not make the student guess what is required.
2. Test only essential information; never test unimportant details.
3. Use closely related materials throughout an item. If students can divide the alternatives into distinct groups, the item is reduced to several multiple-choice items with few alternatives, and the possibility of guessing is distinctly increased.
4. Make all alternatives plausible responses to each element in the first column, wherever possible, to minimize guessing by elimination.
5. Use the working language of the student, wherever possible. By reducing language barriers, both the validity and reliability of the test will be improved.
6. Arrange the alternatives in some sensible order. An alphabetical arrangement is common.
7. If alternatives are not to be used more than once, provide three or four extra ones to reduce guessing.

Effective Item Writing

This is one of the instructor's most difficult tasks. Besides requiring considerable time and effort, this task demands a mastery of the subject, an ability to write clearly, and an ability to visualize realistic situations for use in developing problems. Because effective items are difficult to secure, a semipermanent record of items which have been used is desirable. The record of the item, along with the analysis compiled about the item, can be entered on test item cards. When sufficient item cards are assembled, a test item pool is created. As long as precautions are taken to safeguard the security of items in the pool, the existence of the pool lightens the instructor's burden of continuously preparing new items.

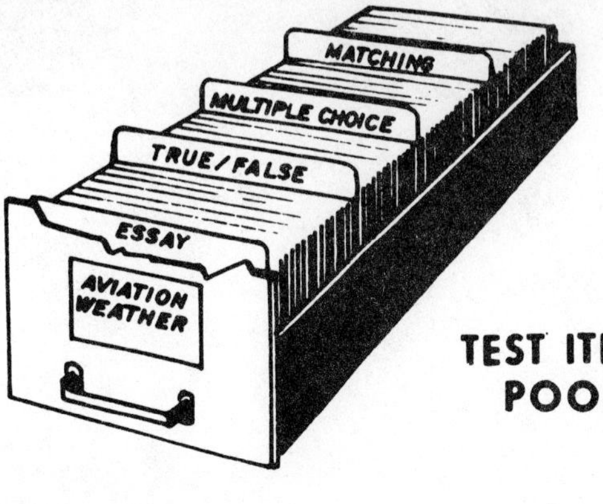

TEST ITEM POOL

Principles to Follow

Regardless of item type or form, the following principles should be followed in writing new items or in reviewing existing items:

1. Each item should test a concept or idea that is important for the student to know, understand, or be able to apply.
2. Each item must be stated so that everyone who is competent in the subject-matter area would agree on the correct response.
3. Each item should be stated in the working language of the student.
4. The wording of the item should be simple, direct, and free of ambiguity. The wording should be edited for brevity. Unnecessary words merely delay the student.
5. Sketches, diagrams, or pictures should be included when they are necessary for the student to visualize the problem correctly or when they will add realism.
6. The item as a whole should present a problem which demands knowledge of the subject or course. No item which can be responded to on the basis of general knowledge should be included in an achievement test.

PERFORMANCE TESTS

If students demonstrate the ability to perform selected parts of a skill for which they are being trained, it is assumed that they will be able to perform the entire skill. In other words, performance testing is a sampling process. It should be a carefully selected part of a complete doing process typical of the skill for which training is being given.

Uses of Performance Testing

This method of evaluation is particularly suited to the measurement of student abilities in *doing*, either mental or physical. Performance testing is desirable for evaluating training that involves an operation, a procedure, or a process.

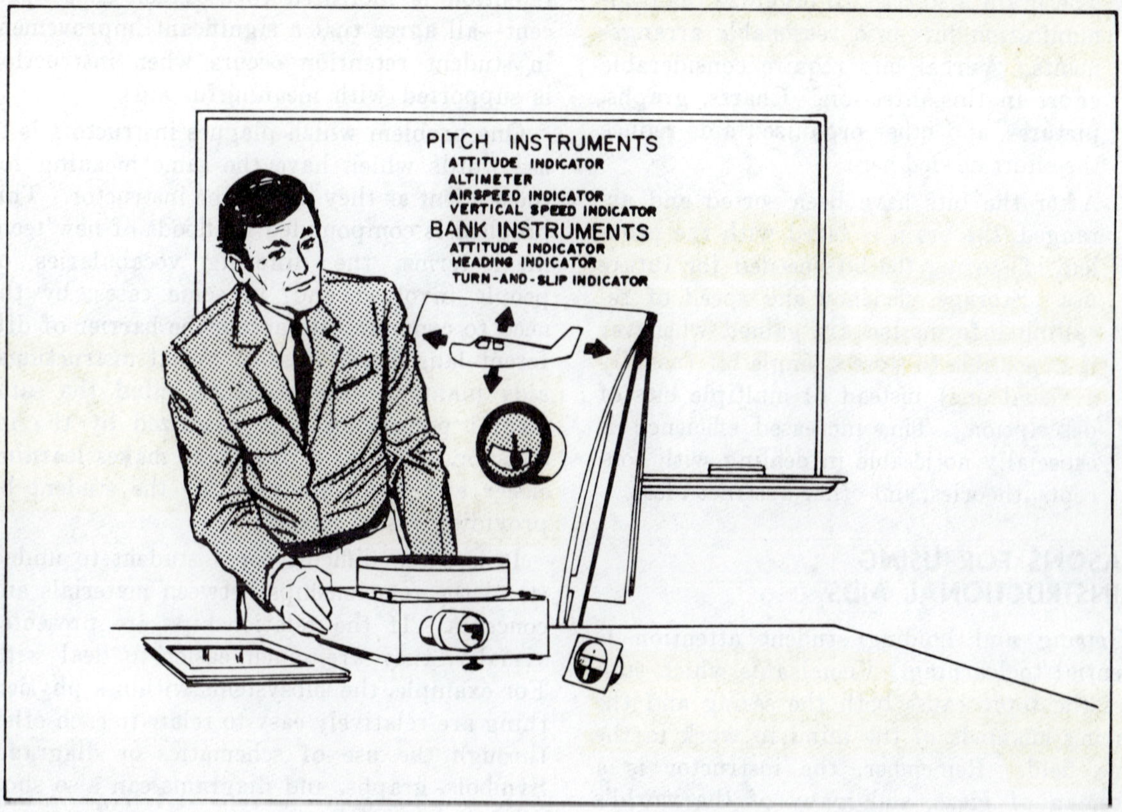

CHAPTER VIII. INSTRUCTIONAL AIDS

Instructors can use instructional aids to improve communication between themselves and their students. Instructional aids are defined by the Department of Audiovisual Instruction of the National Education Association as "Devices which assist an instructor in the teaching-learning process by simply presenting supporting or supplementary material, usually intermittently. They are not self-supporting." They may be either sight or sound devices, or a combination of both. The key factor is that aids support, supplement, or reinforce.

While instructors may become involved in actually preparing aids, their main function is to plan the use of aids so that they help to achieve maximum learning. They must understand some of the theory behind aids, some reasons for using aids, and some guidelines for their use. They should also be familiar with different types of aids.

THEORY BEHIND USE OF INSTRUCTIONAL AIDS

For many years, experts have been theorizing concerning the activity which takes place within the human brain during the communicative process. There is much difference of opinion and little positive agreement concerning this matter. However, there is general agreement about three items of theory that seem pertinent to an understanding of the use of aids.

1. During the communicative process, the mind acts as a filter. As great numbers of verbal or visual bits of communication are received, the brain works to sort out the important items from the routine descriptive or supportive items. Therefore, learning can probably be furthered by helping to identify the important bits. One way to do this is by usiıg aids to emphasize them.

2. The brain also tries to organize the communication bits into reasonable arrangements. Verbal bits require considerable effort in this direction. Charts, graphs, pictures, and other organized aids reduce the effort needed here.
3. After the bits have been sorted and arranged, the brain is faced with the problem of storing the bits needed for future use. Storage efficiency and speed of recalling information are gained whenever it is possible to store a single bit (usually a visual one) instead of multiple bits of description. This increased efficiency is especially noticeable in dealing with concepts, theories, and other abstract ideas.

REASONS FOR USING INSTRUCTIONAL AIDS

Getting and holding student attention is essential to learning. Visual aids which support the topic cause both the seeing and the hearing channels of the mind to work in the same field. Remember, the instructor is a salesman of ideas, and many of the world's sales techniques for getting the attention of clients are worth considering. One caution—get student attention on the subject, not just on a distracting gimmick.

An important goal of all instruction is for the student to retain as much of the instruction as possible, and certainly the key points. Many studies have been made to determine the effect of instructional aids on learning retention. While the specific results of the studies vary greatly—from modest results in which retention increases 10 to 15 percent, to (perhaps overly optimistic) studies in which retention is increased to as much as 80 percent—all agree that a significant improvement in student retention occurs when instruction is supported with meaningful aids.

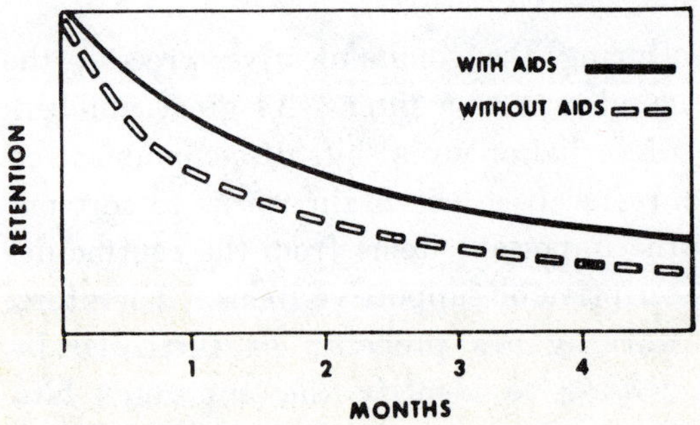

INSTRUCTIONAL AIDS IMPROVE RETENTION

One problem which plagues instructors is to use words which have the same meaning for the student as they do for the instructor. This problem is compounded by floods of new technical terms, the differing vocabularies of people involved and, in some cases, by the need to communicate across the barrier of different languages. Use of visual instructional aids plants in the student's mind the same mental picture that is visualized by the instructor. The good instructor makes learning easier and more accurate for the student by providing visual images.

It is often difficult for a student to understand the relationships between materials and concepts. If the relationships are presented visually, they are much easier to deal with. For example, the subsystems within a physical thing are relatively easy to relate to each other through the use of schematics or diagrams. Symbols, graphs, and diagrams can also show relationships of location, size, time, frequency, and value. By symbolizing the factors involved, it is even possible to visualize relationships between abstracts.

Instructors are frequently asked to teach more and more in less and less time. Instructional aids can help them do this. Instead of using many words to describe a sound, object, or function, the instructor plays a recording of the sound and shows a picture of the object or function. Consequently, the student learns faster and more accurately.

GUIDELINES FOR THE USE OF INSTRUCTIONAL AIDS

The use of any instructional aid must be planned on its ability to support a specific point in a lesson. A simple four-step procedure can be used to determine if and where aids are necessary.

1. First, clearly establish the lesson objective. Be certain of what is to be communicated. This is the most important step.
2. Next, gather the necessary data by researching for support material.

3. Organize the material into an outline or a lesson plan. The plan should include all key points that are to be presented.

4. Finally, and only after completing the previous steps, determine what ideas should be supported with instructional aids. The aids should be concentrated on the key points. Aids are often appropriate when long segments of technical description are necessary, when a point is complex and difficult to put into words, when instructors find themselves forming mental visual images, or when students are puzzled by an explanation or description.

Aids should be simple and compatible with the learning outcomes to be achieved. Obviously, an explanation of elaborate equipment may require detailed schematics, but less complex equipment may lend itself to only basic shapes or line figures. Since aids are normally used in conjunction with a verbal presentation, words on the aid should be kept to a minimum. The instructor should avoid the temptation to use the aids as a crutch. In many cases, visual symbols and slogans can replace extended wordage. The tendency toward unnecessarily distracting artwork should be avoided.

Aids have no value in the learning process if they cannot be heard or seen. Recordings of sounds and speeches should be tested for correct volume and quality in the actual environment in which they will be used. Visual aids must be visible to the entire class. All lettering and illustrations must be large enough to be seen easily by the students farthest from the aids. Colors, when used, should be in clear contrast and easily visible. Numerous rules of thumb can be applied to visibility. The surest and most successful rule is actually to test the aids in the environment in which they will be used.

The effectiveness of aids can be improved by proper sequencing. Frequently, good organization and natural patterns of logic dictate the sequence. Sequencing can be made relatively simple by using acetate overlays on transparencies, stripping techniques on charts and chalkboards, and by imaginative use of felt and magnetic boards. Sequencing can be emphasized and made clearer by the use of contrasting colors.

The effectiveness of aids and the ease of their preparation can be increased by initially planning them in rough draft form. Revisions and alterations are easier to make at that time than after their completion. The rough draft should be carefully checked for accuracy of information, grammar, spelling, basic balance, clarity, and simplicity.

TYPES OF INSTRUCTIONAL AIDS

Some of the most common aids are chalkboards, projected materials, models, charts, and maps.

Chalkboard

The chalkboard is one of the most widely used tools of learning. Its versatility and effectiveness make it a valuable aid to most types of instruction. Two characteristics of the chalkboard are significant: (1) the material presented on it can be erased, allowing the surface to be used again and again; and (2) the chalkboard serves as an excellent medium for joint student-instructor activity in the classroom.

The following practices are fundamental in the use of the chalkboard:

1. Keep the chalkboard clean.
2. Erase all irrelevant material.
3. Keep chalk, erasers, cleaning cloths, rulers, and other aids readily available to avoid interruption of the presentation.
4. Organize and practice chalkboard presentation in advance.
5. Write or draw large enough for everyone in the group to see.
6. Do not overcrowd the chalkboard. Leave a margin around the material and sufficient space between lines of copy.
7. Present material simply and briefly.
8. If necessary, use the ruler, compass, or other devices in making drawings.
9. Use colored chalk for emphasis.
10. Underline statements for emphasis.
11. Use a pointer when discussing chalkboard materials.

12. Stand to the side of the material being presented, so that the entire class will have an unobstructed view.

Models

A model is a realistic copy of an actual piece of equipment. Models are not necessarily the same size as the equipment they represent, nor are they necessarily workable. Models can be used effectively in explaining operating principles of various types of equipment. They are especially adaptable to small group discussions in which students are encouraged to ask questions.

A model is even more effective if it works like the original, and if it can be taken apart and reassembled. With the display of an operating model, the students can observe how each part works in relation to the other parts. When the instructor points to each part of the model while explaining these relationships, the students can better understand the mechanical principles involved. As instructional aids, models are usually more practical than originals, because they are lightweight and easily moved.

Charts

Illustrations, graphs, statistics, diagrams, and other visual material can often be displayed in chart form. The most important factor is the chart's content. Charts may be a series of single stiff cardboards or a series of thin flexible flip charts. The material should be displayed in a clear, easily understood format. The location and handling of charts should be carefully planned and rehearsed.

Projected Material

This grouping includes motion pictures, filmstrips, slides of several sizes, transparencies for overhead projection, and materials used in the opaque projector. The essential factor governing their use is that the content must support the lesson.

Motion pictures can bring many realistic situations to the classroom. Usually, they emphasize key points with no unnecessary details. The combination of motion and sound appeals to students. The availability of packaged lesson segments is attractive to instructors. Care should be exercised to assure that the lesson is being supported, not supplanted. Films should be previewed and summarized before use.

Use of other projected materials can be flexible. The sequence of filmstrips is predetermined, but it is possible and sometimes wise to show only selected frames. Slides, transparencies, and opaque items can be placed in any sequence. Use of projected materials requires careful planning and rehearsal by the instructor for adjusting equipment, lighting, and timing.

FUTURE DEVELOPMENTS

Recent years have seen an explosion of new materials and techniques in the field of instructional aids. They present many new opportunities for instructors, but may also be confusing due to their very abundance and variety. The instructor must keep in mind the teaching goals to be achieved and be selectively receptive to new possibilities. Recent developments include improved projectors, 8mm. film loops, new slide production cameras, simplified techniques and equipment for making transparencies, rear screen projection, improved adhesives, and many more materials for models.

Effective instructors must keep abreast of both new materials and the potential uses for them. They should read extensively in professional journals about these new instructional materials. Above all, they should use creativity and imagination. There is always a better way to accomplish a learning outcome.

SECTION TWO
THE FLIGHT INSTRUCTOR

CHAPTER IX. FLIGHT INSTRUCTOR CHARACTERISTICS AND RESPONSIBILITES

PROFESSIONALISM

As stated in the preamble of FAR Part 61, the flight instructor is the keystone of the new total operational concept and is responsible for all phases of required training. In light of this statement, it is clear that the flight instructor must be a true "professional."

The flight instructor must be fully qualified as a pilot; however, the instructor's ability must go far beyond this if the requirements of professionalism are to be met. Although the term "professionalism" is widely used, it is rarely defined. In fact, no single definition can encompass all of the qualifications and considertions which must be present before true professionalism can exist.

One noted educator has listed the major considerations and qualifications which must be included:

1. Professionalism exists only when a service is performed for someone, or for the common good.
2. Professionalism is achieved only after extended training and preparation.
3. True performance as a professional is based on study and research.
4. Professionalism presupposes an intellectual requirement. Professionals must be able to reason logically and accurately.
5. Professionalism requires the ability to make good judgmental decisions. Professionals cannot limit their actions and decisions to standard patterns and practice.
6. Professionalism demands a code of ethics. Professionals must be true to themselves, and to those they serve. Anything less than a sincere performance is quickly detected, and immediately destroys their effectiveness.

The flight instructor and prospective flight instructor certificate applicant should carefully consider this list. Attempts to operate as a flight instructor without any one of the qualities listed can only result in poor performance and deficient students. Preparation and performance as a flight instructor with these qualities constantly in mind will command recognition as a professional in the field of flight instruction.

Professionalism also includes a flight instructor's public image. In the past, flight instructors have all too often been willing to accept a less-than-professional status in the public view by relaxing their demeanor, appearance, and approach to their profession. A flight instructor who gives the impression that interest in flight instruction is secondary to interest in other activities cannot retain the reputation of a professional. This does not mean that the part-time instructor cannot be a professional. During the time devoted to flight instruction, however, this individual should present a professional image to the public.

If the role of the flight instructor in the aviation industry is to be upgraded, it must be done through the efforts of flight instructors themselves. The professional flight instructor commands the respect of associates; asks for and deserves higher pay; and, most importantly, delivers more effective flight instruction. The following is a discussion of several basic performance factors which should be considered by a flight instructor who seeks to be a professional.

Sincerity

The professional flight instructor should be straightforward and honest. Attempting to hide some inadequacy behind a smokescreen of unrelated instruction will make it impossible for the instructor to command the interested attention of a student.

Teaching a student pilot is predicated upon acceptance of the flight instructor as a competent, qualified teacher and an expert pilot. Any facade of instructor pretentiousness, whether it be real or mistakenly assumed by the student, will immediately cause a loss of confidence by the student in the instructor, and little learning will be accomplished.

The effectiveness of instructor emphasis on the precepts of safety will be lost if the instructor appears to disregard them; the same applies to the instructor's insistence on precision and accuracy in handling an airplane. The professional flight instructor should be honest in every way.

Acceptance of the Student

The professional flight instructor must accept students as they are, with all their faults and all their problems. The student is a person who wants to learn to fly, and the instructor is a person who is available to help in the learning process. Beginning with this understanding, the professional relationship of the instructor with the student should be based on a mutual acknowledgment that both the student and the instructor are important to each other, and that both are working toward the same objective.

Under no circumstances should the professional instructor do anything which implies degrading the student. Acceptance, rather than ridicule, and support, rather than reproof, will encourage learning, regardless of whether the student is quick to learn or is slow and apprehensive. Criticizing a student pilot who does not learn rapidly is not unlike a doctor reprimanding a patient who does not get well as rapidly as was hoped.

Personal Appearance and Habits

Personal appearance has an important effect on the professional image of the instructor. Today's aviation customers are people who expect their associates to be neat, clean, and appropriately dressed. It is not intended that the flight instructor should assume an attire foreign to the flight environment; however, since the instructor is engaged in a learning situation with professional people, the attire worn should be appropriate to a professional status.

Personal habits have a significant effect on the professional image. The exercise of common courtesy is perhaps the most important of these. A flight instructor who is rude, thoughtless, and inattentive cannot hold the respect of the students, regardless of piloting ability.

Cleanliness of body and breath is important to flight instruction. The airplane cabin is a close, tightly-sealed area, where an instructor and a student work in close proximity, and where even little annoyances provide serious distractions from the learning tasks at hand. Smoking by the instructor may, for example, be most unpleasant and distracting for a nonsmoking student.

The use of alcohol in public, especially around an airport, has a serious impact on the flight instructor's image. The smell of alcohol on an instructor's breath during a period of flight instruction is inexcusable.

Demeanor

The attitude and movements of the flight instructor can contribute much to a professional image. The instructor should avoid erratic movements, distracting speech habits, and capricious changes in mood. The professional image requires development of a calm, thoughtful, and disciplined, but not somber, demeanor.

The instructor should avoid any tendency toward frequently countermanding directions, reacting differently to similar or identical errors at different times, demanding unreasonable performances or progress, or criticizing a student unfairly.

A forbidding or imperious demeanor is as much to be avoided as is an air of flippancy. Effective instruction is best fostered by a calm, pleasant, thoughtful demeanor which puts the student at ease, and maintains the instructor's personal image of competence and genuine interest in the student's learning tasks.

Safety Practices and Accident Prevention

The flying habits of the flight instructor, both during flight instruction and as observed by students when conducting other pilot operations, have a vital effect on safety. Students consider their flight instructor to be a paragon of flying proficiency whose flying habits they, consciously or unconsciously, attempt to imitate. The instructor's advocacy and description of safety practices mean little to a student if the instructor is observed to violate them.

For this reason, a flight instructor must meticulously observe the safety practices taught the students. A good example is the use of a checklist before takeoff. If a student sees the instructor start an airplane and take off without referring to a checklist, no amount of instruction in the use of a checklist will assure that the student will use one conscientiously when solo flight operations begin.

A flight instructor must carefully observe all regulations and recognized safety practices during all flight operations if a professional image is to be maintained. An instructor who is observed to fly with apparent disregard for loading limitations or weather minimums creates an image of irresponsibility which many hours of conscientious flight instruction cannot correct.

Habitual observance of regulations, safety precautions, and the precepts of courtesy will enhance the instructor's image of professionalism. Further and more important, such habits make the instructor more effective by developing the same habits in the students.

Perhaps the most productive action a flight instructor can take to enhance flying safety is to actively participate in the Federal Aviation Administration Accident Prevention Program. Although all segments of the FAA are vitally concerned with the program, the flight instructor will be involved chiefly through the General Aviation District Office Accident Prevention Specialist (APS). The program's objective is to improve safety in general aviation by: (1) improving attitudes, (2) increasing knowledge and proficiency through education, and (3) reducing environmental hazards. The GADO Accident Prevention Specialist utilizes the following programs to reduce hazards in these three areas:

1. *Safety meetings.* These meetings are very effective in updating knowledge and proficiency. In planning a safety meeting, the flight instructor should coordinate with the FAA Accident Prevention Specialist regarding time, place, and subject matter. The District Office has the necessary equipment and numerous audio-visual presentations available on various safety subjects, which are appropriate for such meetings.
2. *Accident Prevention Counselors.* These individuals are selected by the District Office Chief and are well-known and highly respected members of the community. They generally are pilots, flight instructors, or aviation maintenance technicians; however, this is not a prerequisite for selection. Counselors are volunteers who are willing to devote time, energy, and thought toward the objective of solving aviation safety problems in their community. To accomplish this

objective, these individuals: (1) counsel airmen who commit unsafe acts; (2) offer assistance and provide information to pilots and aviation organizations in the community in establishing safety programs; (3) make recommendations on matters designed to reduce aviation accidents; (4) publicize accident prevention program activities and accomplishments; (5) encourage the scheduling and assist in the conduct of local aviation safety education meetings, clinics, and seminars; (6) advise the GADO Accident Prevention Specialist when unable to remedy a hazardous situation; and (7) promote and conduct proficiency flights.

3. *Flight Assists.* The Accident Prevention Specialist uses information contained in ATC Flight Assist Reports to help pilots fly safely. No actual or implied punitive action is involved in the utilization of these reports.

The flight instructor must go beyond the requirements of developing technically proficient students who are knowledgeable in the areas of their equipment, flight procedures, and maneuvers. The flight instructor must not only teach students to know their own and their equipment's limitations, but must also teach them to be guided by those limitations. In brief, the flight instructor must make a strenuous effort to *develop good judgment* on the part of the students.

Proper Language

In flight instruction, as in other professional activities, the use of profanity and obscene language leads to distrust or, at best, to a lack of complete confidence. To many people, such language is actually objectionable to the point of being painful. The professional flight instructor must speak normally and without inhibitions, but must not develop the inability to speak positively and descriptively without excesses of language.

The beginning student pilot is entering a realm of new concepts and experiences and is also encountering new terms and phrases which are often confusing. Words such as "traffic," "stall," "elevator," and "lift" are familiar but are given entirely new meanings. Coined words, such as "vortac," "unicom," and "pireps" cause further difficulty. Phrases such as "clear the area," "monitor ATIS," or "lower the pitch attitude" are completely incomprehensible.

The flight instructor does, and properly should, use these terms and phrases during instruction. Although the language of aviation is new and strange, it is part of the new world of flying which the beginning student is eager to learn about and adopt. Difficulty arises, however, when the instructor introduces these new expressions for the first time during an inflight situation, which may be difficult for the student under the best of circumstances.

At the beginning of the student's flight training, and before each flight lesson during early dual instruction, the flight instructor should carefully define the terms and phrases which will be used during the forthcoming lesson. The instructor should then be careful to limit instruction to those terms and phrases, unless the exact meaning and intent of any new expression are explained immediately.

Serious student errors and confusion involving unfamiliar terms and phrases result from the use of the colloquial expressions of aviation, which are rarely specific, and have endless variations. Instructing the student to "give it the needle," "throw the cobs to it," or "firewall it" when it is intended that the throttle should be opened for takeoff may be picturesque and brighten the instruction given; however, the use of such an expression can have serious consequences if one is used for the first time in a critical flight situation. There is the apocryphal story of the pilot of a large airplane faced with an emergency pullup who shouted "takeoff power" to the flight engineer who immediately closed all four throttles.

Self-Improvement

Professional flight instructors must never become complacent or satisfied with their own qualifications and ability. They should be constantly active and alert for ways to improve their qualifications, effectiveness, and the services they provide to students. Flight instructors are considered authorities on aeronautical matters and are the experts to whom many pilots refer questions concerning regulations, requirements, and new operating techniques. They have the opportunity and responsibility of introducing new procedures and techniques through their students and through certificated pilots with whom they come in contact.

There are many means of self-improvement available to flight instructors. Properly organized pilot safety symposiums and flight training clinics are valuable sources of refresher training and of opportunities to exchange information with instructors from other areas. Aviation periodicals, government publications, and technical issuances from the aviation industry are sources of valuable information for flight instructors.

For a professional performance as a flight instructor, it is essential that the instructor maintain current copies of the *Federal Aviation Regulations* that are pertinent to pilot qualification and certification, an *Airman's Information Manual*, current FAA *Flight Test Guides*, and pilot training manuals. A flight instructor who is not completely familiar with current pilot certification and rating requirements cannot do a competent job of flight instruction. The FAA Advisory Circular Checklist and Status of Federal Aviation Regulations, AC 00–2, includes a number of government publications which the instructor should consider for inclusion in a library. The Appendix gives instructions for ordering the checklist and also for obtaining other government reference and training materials. In addition to government publications, a number of excellent handbooks and other reference materials are available from commercial publishers. Also, many public and institutional libraries have excellent resource material on educational psychology, teaching methods, testing, and aviation-related subjects.

HELPING STUDENT PILOTS LEARN

Learning to fly should be an enjoyable experience. By making each lesson a pleasurable experience for the student, the flight instructor can maintain a high level of student motivation. This does not mean the instructor must make things easy for the student or sacrifice standards of performance to please the student. The student will experience pleasure from a learning task well done or from successfully meeting the challenge of a difficult operation.

The idea that people must be led to learning by making it easy has no basis in fact. People are not always attracted to something which is pleasant and easy. Actually, they devote more effort to things which bring rewards, such as self-enhancement and personal satisfaction. People want to feel capable; they are proud of difficult achievements.

Learning to fly should be interesting. Knowing the objective of each period of instruction gives meaning and interest to the instructor's and student's efforts. Not knowing the objective involved leads to confusion, disinterest, and uneasiness on the part of the student.

Learning to fly should provide an opportunity for exploration and experimentation for students. Students should be allowed time to explore and evaluate the various elements of each maneuver or operation presented and thereby discover their own capabilities and acquire self-confidence. This can also be fostered by using alternative presentations for different students.

Learning to fly should be a habit-building period during which students devote their attention, memory, and judgment to the development of correct habit patterns. Any goal other than a desire to learn the right way makes students impatient of the instruction and practice they need and should be trying to obtain. The instructor should keep this goal before the students by example and by a logical presentation of learning tasks.

As was stated at the beginning of this chapter, flight instructors have full responsibility for all phases of required training. To meet this responsibility, flight instructors must be clear regarding their objectives. After the objectives have been established, teaching methods and activities must be organized to best achieve them.

Instructors must take specific steps if student learning is to be effectively fostered. They must (1) devise a plan of action, (2) create a positive student-instructor relationship, (3) present information and guidance effectively, (4) transfer responsibility to the student as learning occurs, and (5) evaluate student learning and thereby their own teaching effectiveness. While these distinct factors involved in instruction are not apparent to the student during learning, the disregard of any one of them results in a difficult and inefficient learning experience.

Helping the student learn does not mean that the instructor has the responsibility for performing learning tasks which students can do for themselves. This is not effective instruction. The best instructors provide only the information, guidance, and opportunity for student learning, and support their motivation while they are in a learning situation.

Providing Adequate Instruction

The flight instructor must attempt to analyze carefully and correctly the personality, thinking, and ability of each student. No two students are alike, and the same methods of instruction cannot be equally effective for all students. The instructor must talk with a student at some length to learn about the student's background, interests, way of thinking, and temperament. The instructor's methods may change as the student advances through successive stages of training; a gentle introduction must sometimes be followed by strict instruction if progress is to continue in advanced stages.

An instructor who has not correctly analyzed a student may soon find that the instruction is not producing the desired results. This could mean, for example, that the instructor has analyzed as a slow thinker a student who is actually a quick thinker but is hesitant to act. Such a student may fail to act at the proper time due to lack of self-confidence, even though the situation is correctly understood. In this case, the correction would obviously be instruction directed toward developing student self-confidence, rather than drill on flight fundamentals.

The slow student requires instructional methods which combine tact, keen perception, and delicate handling. If such a student receives too much help and encouragement, a feeling of incompetence may develop. Too much criticism may completely subdue a timid person, whereas brisk instruction may force a more diligent application to the learning task.

A student whose slow progress is due to discouragement and a lack of confidence should be assigned "subgoals" which can be attained more easily than the normal learning goals. For this purpose, complex flight maneuvers can be separated into their elements, and each element practiced until an acceptable performance is achieved before the whole maneuver or operation is attempted. As an example, instruction in turns across a road may begin with consideration at first for headings only, and the problems of altitude control, drift correction, and coordination can be introduced separately, one at a time. As the student gains confidence and ability, goals should be increased in difficulty until progress is normal.

Apt students can also create problems. Because they make few mistakes, they may assume that the correction of errors is unimportant. Such overconfidence soon results in faulty performance. For such students, a good instructor will constantly raise the standard of performance for each lesson, demanding greater effort. Individuals learn when they are aware of their errors. Students who are permitted to complete every flight lesson without corrections and guidance will not retain what they have practiced as well as those students who have their attention constantly directed to the analysis of their performance. This does not mean that deficiencies must be invented for their benefit, because unfair criticism immediately destroys a student's confidence in the instructor.

The demands on an instructor to serve as a practical psychologist are much greater than is generally realized. An instructor can meet this responsibility only through a careful analysis of the students and through a continuing deep interest in them.

Demanding Adequate Standard of Performance

Flight instructors must continuously evaluate their own effectiveness and the standard of learning and performance achieved by the students. The desire to maintain pleasant personal relationships with the students must not cause the acceptance of a slow rate of learning or a low level of flight performance. It is a fallacy to believe that accepting lower standards to please a student will affect a genuine improvement in the student-instructor relationship. Reasonable standards strictly enforced are not resented by an earnest student.

Flight instructors fail to provide competent instruction when they permit their students to get by with a substandard performance, or without learning thoroughly some item of knowledge pertinent to safe piloting. More importantly such deficiencies may in themselves allow hazardous inadequacies in the students' later piloting performance.

Emphasizing the "Positive"

Flight instructors have a tremendous influence on their students' "image" of aviation in general and piloting in particular. The way flight instructors conduct themselves, the attitudes they display, and the manner in which they develop their instruction all contribute to the formation of either *positive* or *negative* impressions by their students. Flight instructor success depends, in large measure, on the ability to frame instructions so that students develop a positive image of flying.

In Chapter I, it was emphasized that negative self-concepts inhibit the perceptual process, that fear adversely affects the students' perceptions, that threat limits their ability to perceive, and that negative motivations are not as effective as positive motivations. A knowledge of these factors, which have such a profound effect on the students' ability to absorb instruction is not enough. Instructors must be constantly aware of these and other "negativisms" and not allow them to creep into their instruction.

Consider how the following not-too-exaggerated first flight lesson might impress a new student pilot without previous experience in aviation:

1. An exhaustive indoctrination in preflight procedures, with emphasis on the extreme precautions which must be taken before every flight, because mechanical failures in flight are often disastrous.
2. Instruction in the extreme care which must be taken in taxiing an airplane, because "if you go too fast, it's likely to get away from you."
3. A series of stalls, because "this is how so many people lose their lives in airplanes." (The side effect of this performance on the first lesson is likely to be airsickness.)
4. A series of simulated forced landings, because one should always be prepared to cope with an engine failure.

These are a series of new experiences which might make the new student wonder whether learning to fly is a good idea or not.

For contrast, one might consider a first flight lesson in which the preflight inspection is presented to familiarize the student with the airplane and its components, and the flight consists of a perfectly normal flight to a nearby airport and return. Following the flight, the instructor can call the student's attention to the ease with which the trip was made in comparison with other modes of transportation, and the fact that no critical incidents were encountered or expected.

This by no means proposes that preflight inspections, stalls, and emergency procedures should be omitted from training. It only illustrates the "positive" approach, in which the student is not overwhelmed with the critical possibilities of aviation before having an opportunity to see its potential and pleasurable features. The introduction of emergency procedures after the student has developed an acquaintance with normal operations is not so likely to be discouraging and frightening, or to retard learning by the imposition of fear.

There is no creed in aviation which demands that students must suffer as part of their flight instruction. This has often been the case because of the unthinking use of "negative" explanations and motivations for all flight operations. Every effort should be made to assure that flight instruction is given under the most favorable conditions.

There is the unfamiliar vibration, the strange noises, the eerie sensations due to "g" loads, or the "woozy" feeling in the stomach. Instructors, to be effective, cannot ignore the existence of these negative factors, nor should they ridicule students who are adversely affected by them. Rather, these negativisms must be overcome by positive instruction.

An instructor may explain to a student that a flight maneuver or procedure must be accomplished in a certain manner. To perform it otherwise, the instructor points out, is to flirt with disaster or to suffer serious consequences. Justifications such as these may be very convenient, and the instructor may consider such negative justifications sufficiently dramatic to assure that the point is committed to memory. The final test, however, must be whether the stated reasons contribute to the learning situation. With very few exceptions, the results which can be expected should be very apparent. *Negative teaching generally results in negative learning.*

Most new flight instructors tend to adopt those teaching methods used when they were students. These methods may or may not have been good. The fact that one has learned to fly under one system of instruction does not mean that this is necessarily the best way it can be done, regardless of the respect one retains for the ability of an old instructor. Some students learn to fly in spite of their instruction, rather than because of it.

EVALUATION OF STUDENT PILOTING ABILITY

This is one of the basic elements of flight instruction. In flight instruction, the instructor determines by oral quizzing that the student understands the procedure or maneuver to be learned, demonstrates its performance, allows the student to try it out and practice it under direction, and then evaluates student accomplishment by observing performance.

Evaluation of demonstrated ability during flight instruction must be based upon established standards of performance, suitably modified to apply to the student's experience and stage of development as a pilot. The evaluation, to be meaningful to the instructor, must consider the student's mastery of the elements involved in the maneuver, rather than merely the overall performance.

In flight instruction, demonstrations of piloting ability are important for exactly the same purposes as are quizzes. They have additional special significance, however, in being directly applied to the qualification of student pilots for solo and solo cross-country privileges. Also associated with pilot skill evaluations during flight instruction are the stage completion checks conducted in approved flying courses and flight checks for pilot certification flight-test recommendations.

In evaluating student demonstrations of piloting ability, as in quizzing and other instructional processes, it is important for the flight instructor to keep the student informed of progress. This may be done as each procedure or maneuver is completed or during postflight critiques.

Corrections or the explanations of errors in performance should point out the elements in which the deficiencies are believed to have originated and, if possible, appropriate corrective measures should be suggested. Correction of student errors should not include the practice of taking the controls away from the student every time a mistake is made. A student may perform a procedure or maneuver correctly and not fully understand the principles and objectives involved. When this is suspected by the instructor, the student should be required to vary the performance of the maneuver slightly, combine it with other operations, or apply the same elements to the performance of other maneuvers. A student who does not understand the principles involved will probably not be able to do this successfully.

THE FLIGHT INSTRUCTOR AS A PRACTICAL PSYCHOLOGIST

Flight instructors must be able to evaluate student personality if they are to use appropriate techniques in the presentation of instruction. While it is obviously impossible for every flight instructor to be an accomplished psychologist, there are a number of considerations which will assist in learning to analyze students before and during each lesson.

Anxiety

Anxiety is probably the most significant psychological factor affecting flight instruction. This is true because flying is a potentially threatening experience for persons who are not accustomed to being off the ground. The fear of falling is universal in human beings.

Anxiety is described by Webster as "a state of mental uneasiness arising from fear. . . ." It results from the fear of anything, real or imagined, which threatens the person who experiences it, and may have a potent effect on actions and on ability to learn from perceptions.

The responses to anxiety vary greatly. They range from a hesitancy to act to the impulse to "do something even if it's wrong!" Some persons affected by anxiety will react appropriately, adequately, and more rapidly than they would in the absence of threat. Many persons, on the other hand, may be frozen in place and incapable of doing anything to correct the situation which has caused their anxiety. Others may do things without rational thought or reason.

Both normal and abnormal reactions to anxiety are of concern to the flight instructor; the normal, because they indicate a need for special instruction to relieve the anxiety which causes them, and the abnormal, because they may be evidence of deep-seated trouble.

Anxiety can be countered by reinforcing students' enjoyment of flying, and by teaching them to cope with their fears. An effective technique is to treat fears as a normal reaction, rather than ignoring them.

Anxiety, for student pilots, is usually associated with the performance of certain flight maneuvers and operations. Instructors should introduce such maneuvers with care, so that students know what to expect, and what their reactions should be. When introducing stalls, for example, instructors should first explain the aerodynamic effects involved, and then carefully describe the sensations to be expected and the responses demanded of the pilot.

Student anxieties can be minimized throughout training by emphasizing the benefits and pleasurable experiences which can be derived from flying, rather than by continuously citing the unhappy consequences of faulty performances. Safe flying practices should be presented as conducive to satisfying, efficient, uninterrupted operations, rather than as necessary only to prevent catastrophe.

Normal Reactions to Stress

When a threat is recognized or imagined, the brain alerts the body. The adrenal gland then pours out hormones which prepare the body to meet the threat, or to retreat from it. The heart rate quickens, certain blood vessels constrict to divert blood to the organs which will need it, and other changes take place.

Normal individuals begin to respond rapidly and exactly, within the limits of their experience and training. Many responses are automatic, which points up the need for proper training in emergency operations prior to an actual emergency. The affected individual thinks rapidly, acts rapidly, and is extremely sensitive to all aspects of the surroundings.

Abnormal Reactions to Stress

With certain persons the same bodily reaction to stress does not produce actions which we regard as normal. With them, response to anxiety or stress may be completely absent or at least inadequate. Their responses may be random or illogical, or they may be more than is called for by the situation.

Flight instructors are the only persons, during flight instruction, who observe students when the "pressure is on." They are, therefore, the only individuals in a position to differentiate between potentially "safe" and "unsafe" pilots psychologically.

Flight instructors may accept the following student reactions as indicative of abnormal reactions to stress. None of them provides an absolute indication, but the presence of any of them under conditions of stress is reason for careful instructor evaluation.

1. Autonomic responses such as sweating (especially in the palms), rapid heart rate, paleness, etc.
2. Inappropriate reactions, such as extreme overcooperation, painstaking self-control, inappropriate laughter or singing, very rapid changes in emotions, and motion sickness under conditions of stress.
3. Marked changes in mood on different lessons, such as excellent morale followed by deep depression.
4. Severe anger at the flight instructor, service personnel, or others.

Flight instructors who find themselves involved with psychologically abnormal students may feel that they are not meeting the students' needs. They may believe that student actions are intended to be insulting, or more often, find student actions simply confusing. In difficult situations of this sort, flight instructors must carefully examine student responses and their own responses to the students. These responses may be the normal products of a complex learning situation, but they can also be indicative of psychological abnormalities which will inhibit learning, or be potentially very hazardous to future piloting operations.

Flight Instructor Actions Regarding Seriously Abnormal Students

A flight instructor who believes, after a careful consideration of all available evidence, that a student may be suffering from a serious psychological abnormality has a legal and moral responsibility to refrain from certifying that student to be a competent pilot.

The flight instructor's primary legal responsibility concerns the decision whether to certify the student to be competent for solo flight operations, or to execute a flight-test recommendation leading to certification as a pilot. If, after consultation, the instructor believes that the student suffers a serious psychological deficiency, such authorizations and recommendations must not be signed.

Flight instructors have the personal responsibility of assuring that such a person does not continue flight training or become certificated as a pilot. To accomplish this, the following steps are available:

1. If an instructor believes that a student may have a disqualifying psychological defect, arrangements should be made for another instructor, who is not acquainted with the student, to conduct an evaluation flight. After the flight, the two instructors should confer to determine whether they agree that further investigation or action is justified.
2. An informal discussion should be initiated with the local General Aviation or Flight Standards District Office, suggesting that the student may be able to meet the skill standards, but may be unsafe psychologically. This action should be taken as soon as a question arises regarding the student's fitness. It should not be delayed until the student feels competent to solo.
3. A discussion should be held with a local Aviation Medical Examiner, preferably the one who issued the student's medical certificate, to obtain advice and to decide on the possibility of further examination of the student.

STUDENT PILOT SUPERVISION AND SURVEILLANCE

Flight instructors have the moral obligation to provide guidance and restraint with respect to the solo operations of their students. This applies to instructors' observations of unsafe or inept operations by pilots who are not aware they are being observed, as well as pilots who have requested an instructor's evaluation or guidance. In the case of an observed unsatisfactory performance, it is the instructor's responsibility to try to correct it by the most reasonable and effective means. If unable to correct the situation by personal contacts and good advice, the instructor should report the situation which has caused the observed deficiencies to an Accident Prevention Counselor or a GADO Accident Prevention Specialist.

FLIGHT INSTRUCTOR ENDORSEMENTS

The authority and responsibility for endorsing student pilot certificates and logbooks for solo and solo cross-country flight privileges are a very important flight instructor prerogative. The rules covering these endorsements are contained in Part 61 of the Federal Aviation Regulations and are further explained in Advisory Circular 61–65, Part 61 Certification: Pilot and Flight Instructors. Failure to ascertain that a student pilot meets the requirements of regulations prior to making endorsements is a serious deficiency in performance, for which a flight instructor is held accountable. Providing a solo endorsement for a student pilot who is not fully prepared to accept the responsibility for solo flight operations is also a breach of faith with the student concerned.

Caution should be exercised regarding student pilots who seek to "learn to fly" from pilots who are not certificated flight instructors. This may not be illegal, but the accumulation of flying experience in this manner does not relieve the flight instructor of responsibility for determining that the student has met all the requirements of the regulations.

FLIGHT TEST RECOMMENDATIONS

Provision is made on all private and commercial pilot certificate application forms for the written recommendation of the flight instructor who has prepared the applicant for the flight test involved. The signing of this recommendation imposes a serious responsibility on the flight instructor.

A flight instructor who is asked to execute a flight test recommendation for a pilot certificate applicant should require the applicant to thoroughly demonstrate qualifications for the flight test sought. This demonstration should in no instance be less than the complete test procedure prescribed in the pertinent *FAA Flight Test Guide.*

A flight test recommendation based on anything less risks the presentation of an applicant who may be totally unprepared for some part of the official flight test. In such an event, the flight instructor is logically held accountable for a deficient instructional performance. This risk is especially great in signing recommendations for applicants who have not been trained by the instructor involved. Federal Aviation Regulations require a minimum of 3 hours of dual flight test preparation for a private pilot certificate and 10 hours of such instruction for a commercial certificate.

FAA inspectors and designated pilot examiners rely on flight instructor recommendations as evidence of qualification for certification, proof that a review of the subject areas found to be deficient on the airman written test has been given, and assurance that the applicant has had a thorough briefing on the flight test standards and procedures.

It is very unusual for a competently prepared applicant to fail a flight test. Failure by an incompletely prepared applicant usually results in the need for a greater amount of dual instruction than would have been required for adequate preparation for the original test.

AIRPLANE CHECKOUTS

Flight instructors are often called upon to check out certificated pilots in unfamiliar airplanes. In the case of student pilots, such checkouts and certificate endorsements are required for each make and model airplane the student is allowed to operate solo.

With the increase in popularity of high-performance personal airplanes with distinct flight characteristics, complex systems and equipment, competent checkouts are essential for their safe, efficient operation. Checkouts which consist of a demonstration of the ability to take off and land the airplane concerned are no longer adequate.

FAA Advisory Circular 61-9A, *Pilot Transition Courses for Complex Single-Engine and Light Twin-Engine Airplanes*, provides useful guidance for an instructor. All checkouts should, of course, be conducted to the performance standards required by the appropriate FAA *Flight Test Guide* for the grade of certificate held by the pilot involved.

For the conduct of a pilot checkout, it is essential that the flight instructor be fully qualified in the airplane to be used and be thoroughly familiar with its operating procedures, approved flight manual, and operating limitations. An instructor who does not meet the recent flight experience prescribed by regulations for the airplane concerned should not attempt to check out another pilot.

The flight instructor who checks out a pilot in an aircraft for which a type rating is not required by regulations is accepting a major responsibility for the safety of future passengers when certifying the competency of the pilot. Many of these newer small airplanes are comparable in performance and complexity to transport airplanes. For these, the flight instructor's checkout should be at least as thorough as an official type rating flight test.

For the benefit of the pilot concerned, and for the instructor's protection in the case of later question, the flight instructor should record in the pilot's logbook the exact extent of any checkout conducted. This can be done most easily by reference to the appropriate FAA *Flight Test Guide.*

In the event the instructor finds a pilot's performance in the airplane used sufficiently deficient to constitute a hazard, an attempt should be made to influence the pilot to obtain further instruction before continuing operation of the airplane concerned. If this is unsuccessful, and a real hazard is considered to exist, it is the instructor's responsibility to bring the situation to the attention of the appropriate FAA District Office.

REFRESHER TRAINING

The conduct of refresher training for certificated pilots is not only a responsibility of the flight instructor, but it can also be a profitable opportunity. As stated in FAR Part 61, no person may act as pilot in command unless a flight review has been accomplished within the preceding 24 months.

Effective pilot refresher training must be based on specific objectives and standards if it is to be effective. The objectives should include a thorough checkout appropriate to the grade of certificate and aircraft ratings held, and the standards should be at least those required for the issuance of that grade of certificate. Before beginning any training, the pilot and the instructor should agree fully on these objectives and standards, and, as training progresses, the pilot should be made constantly aware of progress toward their achievement.

FAA Advisory Circular 61-10A, *Refresher Courses for Private and Commercial Pilots*, contains recommended procedures and standards for general pilot refresher courses.

CHAPTER X. TECHNIQUES OF FLIGHT INSTRUCTION

In Chapter V, three teaching methods, the lecture, the guided discussion, and the demonstration-performance, were described in general terms. In this chapter, two teaching techniques useful to the flight instructor are discussed. Both are based on the demonstration-performance teaching method. This chapter also includes a discussion of several obstacles to learning during flight instruction.

THE "TELLING AND DOING" TECHNIQUE IN FLIGHT INSTRUCTION

This technique has been in use for a long time and is very effective in teaching skills. Flight instructors find it valuable in teaching procedures and maneuvers. It is basically the demonstration-performance method and follows the four steps of the teaching process discussed in Chapter IV, except for the first step, PREPARATION. The first step is particularly important in flight instruction because of the new concepts and complexities involved. The flight instructor needs to be well prepared and highly organized if complex ideas and skills are to be taught effectively. The student must be intellectually and psychologically ready for the learning activity. The PREPARATION step is accomplished prior to the flight lesson, by a careful consideration and discussion of objectives, and by a thorough preflight discussion. Steps two, three, and four of the teaching process can be accomplished by "Telling" and "Doing."

Instructor Tells—Instructor Does

This is the second step in the teaching process—PRESENTATION. It is a continuation of preparing the student, which began in the detailed preflight discussion, and now continues by a carefully planned demonstration and accompanying verbal explanation of the procedure or maneuver. It is important that the demonstration conform to the explanation as closely as possible. If a deviation does occur, the instructor should point it out and account for it.

Student Tells—Instructor Does

This is a transition between the second and third steps in the teaching process. It assures the student and the instructor that the explanation and demonstration have been adequate and are thoroughly understood.

Student Tells—Student Does

This is the third step in the teaching process—APPLICATION. This is where learning takes place and where performance habits are formed. If the student has been adequately prepared (first step) and the procedure or maneuver fully explained and demonstrated (second step), meaningful learning will occur. The instructor should be alert during the student's practice to detect any errors in technique and to prevent the formation of erroneous ideas or faulty habits.

Student Does—Instructor Evaluates

This is the fourth step of the teaching process—REVIEW and EVALUATION, in which the instructor reviews what has been covered during the instructional flight and determines to what extent the student has met the objectives outlined during the preflight discussion.

THE INTEGRATED TECHNIQUE OF FLIGHT INSTRUCTION

"Integrated flight instruction" is flight instruction during which students are taught to perform flight maneuvers both by outside visual references and by reference to flight instruments, FROM THE FIRST TIME EACH MANEUVER IS INTRODUCED. No distinction in the pilot's operation of the flight controls is permitted, regardless of whether outside references or instrument indications are used for the performance of the maneuver.

When this training technique is used, instruction in the control of an airplane by outside visual references is "integrated" with instruction in the use of flight instrument indications for the same operations.

Objectives

Integrated flight instruction was introduced on a national scale in 1959, when an amendment to the Civil Air Regulations established certain instruction and competency in the use of flight instruments as prerequisites for the issuance of private pilot certificates. The objective of this training was, and still is, the formation of firm habit patterns for the observance of and reliance on flight instruments from the student's first piloting experience. Such habits have been proved to produce more capable and safer pilots for the efficient operation of today's airplanes. The ability to fly in instrument weather is not the objective of this type of primary training, although it does greatly facilitate later instrument flight training.

Development of Habit Patterns

The continuing observance of and reliance upon flight instruments is essential to the efficient, safe operation of modern high-performance airplanes. The habit of monitoring instruments constantly is difficult to develop after one has become accustomed to relying exclusively on outside references for heading, altitude, airspeed, and attitude information, a procedure which was adequate in most older airplanes.

General aviation accident reports provide ample support for the belief that habitual reference to flight instruments is important to safety. The safety record of pilots who hold instrument ratings is significantly better than that of pilots with comparable flight time who have never received formal instrument flight training.

Student pilots who have been required to perform all normal flight maneuvers by reference to instruments, as well as by outside references, will develop from the start the habit of continuously monitoring their own and the airplane's performance. This habit would be much more difficult for a student to develop after intensive piloting experience without it, as veteran pilots who begin formal training for an instrument rating can readily testify.

The early establishment of proper habits of instrument coverage, instrument interpretation, and aircraft control will be of great assistance to the student pilot in gaining competence in the pilot operation "Maneuvering By Reference To Instruments" which is included in the *Private Pilot Airplane Flight*

Test Guide, AC 61–54A. The habits formed at this time will also give the student a firm foundation for later training for an instrument rating.

Accuracy of Flight Control

During early experiments with the integrated technique of primary flight instruction, it was soon recognized that students trained in this manner are much more precise in their flight maneuvers and operations. This applied to all flight operations, not just when flight by reference to instruments is required.

Notable among students' achievements are better monitoring of power settings and more accurate maintenance of desired headings, altitudes, and airspeeds. As the habit of monitoring their own performance by reference to instruments is developed, students will begin to make corrections without prompting.

The habitual attention to instrument indications leads to improved landings because of more accurate airspeed control, superior cross-country navigation, better coordination, and a generally better overall pilot competency.

Operating Efficiency

As student pilots become more proficient in monitoring and correcting their own flight technique by reference to flight instruments, the performance obtained from an airplane increases noticeably. This is particularly true of modern, high performance airplanes, which are responsive to the use of correct operating airspeeds.

The use of correct power settings and climb speeds and the accurate control of headings during climbs result in a measurable increase in climb performance. The maintenance of headings and altitudes in cruising flight will definitely increase average cruising speeds.

Emergency Capability

The use of integrated flight instruction provides the student with the ability to control an airplane in flight for limited periods if outside references are lost. This ability could save the pilot's life and those of the passengers in an actual emergency.

During the conduct of integrated flight training, the flight instructor must emphasize to the students that their introduction to the use of flight instruments does not prepare them for operations in marginal or instrument weather conditions. The possible consequences, both to themselves and to others, of experiments with flight operations in weather conditions worse than those required for VFR operations before they are instrument rated, should be constantly impressed on the students.

Procedures

The conduct of integrated flight instruction is simple. The use of an airplane equipped with flight instruments and an easily demountable means of simulating instrument flight conditions, such as an extended visor cap, are needed. The student's first briefing on the function of the flight controls should include the instrument indications to be expected, as well as the outside references which should be used to control the attitude of the airplane.

Each new flight maneuver should be introduced using either outside references or instrument indications, as the instructor prefers. The student's visor should then be raised or lowered, whichever is appropriate, and the same maneuver performed by the use of the other set of references. New students, having no inhibitions about instrument flying, rapidly develop the ability to maneuver an airplane equally well by instrument or outside references. They accept naturally the fact that the manipulation of the flight controls is identical, regardless of which references are used to determine the attitude of the airplane. This practice should continue throughout the student's dual instruction for all flight maneuvers except those which require the use of ground references. To fully achieve the demonstrated benefits of this type of training, the use of visual and instrument references must be constantly integrated throughout the training. Failure to do so will lengthen the dual instruction necessary for the student to achieve the competency required for a private pilot certificate.

Precautions

During the conduct of integrated flight instruction, the instructor must be especially vigilant for other air traffic while the student is operating by instrument references. The instructor must guard against having attention diverted to the student's performance for extended periods.

At the same time, the instructor must be sure that the students develop, from the start of their training, the habit of looking for other air traffic at all times when they are not operating under simulated instrument conditions. If students are allowed to believe that the instructor assumes all responsibility for avoiding other traffic, they cannot develop the habit of keeping a constant watch, which is essential to safety. Any observed tendency of a student to enter flight maneuvers without first making a careful check for other possible air traffic must be corrected immediately.

In the earlier stages of training, students may find it easier to perform flight maneuvers by instruments than by outside references. The fact that students can perform better by reference to instruments may cause them to concentrate most of their attention on the instruments, when they should be using outside references. This must not be allowed to continue, since it will cause considerable difficulty later in training while maneuvering by reference to ground objects. This tendency will also limit vigilance for other air traffic. The instructor should carefully observe the student's performance of maneuvers during the early stages of integrated flight instruction to assure that this habit does not develop. If it is detected, the instructor should make the student concentrate on maneuvering by outside references with the gyroscopic instruments caged or covered.

During the conduct of integrated flight instruction, the instructor should make it clear that the use of instruments is being taught to prepare students to accurately monitor their own and their airplane's performance, not to qualify them for IFR operations. The instructor must avoid any indication, by word or action, that the proficiency sought is intended solely for use in difficult weather situations.

Flight Instructor Qualifications

It is essential that a flight instructor be thoroughly familiar with the functions, characteristics, and proper use of all standard flight instruments. It is also the personal responsibility of each flight instructor to maintain familiarity with current pilot training techniques and certification requirements. This may be done by constant use of new periodicals and technical publications, personal contacts with Federal Aviation Administration Inspectors and designated pilot examiners, and by participation in pilot and flight instructor symposiums and clinics. The application of outmoded instructional procedures, or the preparation of student pilots for obsolete certification requirements is inexcusable.

OBSTACLES TO LEARNING DURING FLIGHT INSTRUCTION

Among those obstacles which are common to flight instruction, and which have been recognized as major factors to be considered by flight instructors, are students':

1. Feeling of unfair treatment.
2. Impatience to proceed to more interesting operations.
3. Worry, or lack of interest.
4. Physical discomfort, illness, or fatigue.
5. Apathy, fostered by poor instruction.
6. Anxiety.

Students who believe that their instruction is perfunctory, or that their efforts are not conscientiously considered and evaluated, will not learn well, and their motivation will suffer no matter how intent they are on learning to fly.

UNREASONABLE DEMANDS RETARD LEARNING

Motivation will also decline when a student believes the instructor is making unreasonable demands for performance and progress. The assignment of goals which the student considers difficult, but possible, usually provides a challenge which promotes learning. The assignment of impossible goals discourages the student, diminishes effort, and retards the learning process.

Impatience is a greater deterrent to learning pilot skills than is generally recognized. With a flight student, this may take the form of a desire to make an early solo flight, or to set out on cross-country flights before the basic elements of flight have been learned.

The impatient student fails to understand the need for preliminary training and seeks only the ultimate objective without considering the means necessary to reach it. In flying an airplane, as with every complex human endeavor, it is necessary to master the basics if the whole task is to be performed competently and safely. Student impatience can be corrected by the instructor by presenting the necessary preliminary training one step at a time, with clearly stated goals for each step. The procedures and elements mastered in each step should be clearly identified in demonstrating the performance of the subsequent step.

Impatience can result from instruction keyed to the pace of a slow learner when it is applied to an apt student. It is just as important that a student be advanced to the subsequent step as soon as one goal has been attained, as it is to complete each step before the next one is undertaken. Disinterest grows rapidly when unnecessary repetition and drill are required on operations which have been learned adequately.

Worry or lack of interest has a very detrimental effect on learning. Students who are worried or emotionally upset do not learn well and derive little benefit from any practice performed. Worry or distraction may be due to students' concern about progress in the training course, or may stem from circumstances completely unrelated to their instruction. Significant emotional upsets may be due to personal problems, psychiatric disturbances, or an antipathy for the training concerned or the instructor.

Students' experiences outside their training activities affect their behavior and performance in training; the two cannot be separated. When students begin flight training, they bring with them their interests, enthusiasms, fears, and troubles. The instructor cannot be responsible for these outside diversions, but cannot ignore them because they vitally affect teaching. Instruction must be keyed to the utilization of the interests and enthusiasms students bring with them, and to diverting their attention from their worries and troubles to the learning tasks at hand. This is admittedly difficult, but must be accomplished if learning is to proceed at a normal rate.

Worries and emotional upsets which result from the flight course can be remedied. Such occurrences are usually evidence of inadequacies on the part of the course or of the instructor. The most effective cure is prevention. The instructor must be alert to see that the students understand the objectives of each step of their training, and that they know at the completion of each lesson exactly what their progress and deficiencies have been. Discouragement and emotional upsets are rare when students feel that nothing is being withheld from them or is being neglected in their training.

Physical discomfort, illness, and fatigue will materially slow the rate of learning during both classroom instruction and flight training. Students who are not completely at ease, and whose attention is diverted by discomforts such as the extremes of temperature, poor ventilation, inadequate lighting, or noise and confusion, cannot learn at a normal rate. This is true no matter how diligently they attempt to apply themselves to the learning task.

A minor illness, such as a cold, or a major illness or injury will interfere with the normal rate of learning. This is especially important to the conduct of flight instruction, because most illness adversely affects the acuteness of vision, of hearing, and of feeling which are essential to correct performance.

Airsickness can be a great deterrent to flight instruction. A student who is airsick, or bothered with incipient airsickness, is incapable of learning at a normal rate. There is no sure cure for airsickness, but resistance or immunity can be developed in a relatively short period of time. An instructional flight should be terminated as soon as incipient sickness is experienced. As the student develops immunity, flights can be increased in length until normal flight periods are practicable.

Keeping students interested and occupied during flight is a deterrent to airsickness. They are much less apt to become airsick while operating the controls themselves. Rough air and unexpected abrupt maneuvers tend to increase the chances of airsickness. Tension and apprehension apparently contribute to airsickness and should be avoided.

The detection of student fatigue is important to efficient flight instruction. This is important both in assessing a student's substandard performance early in a lesson, which may be due to inadequate rest, and also in recognizing the deterioration of performance, which results from continuing intensive concentration on a complex task. Once fatigue occurs as the result of application to a learning task, the student should be given a break in instruction and practice. Fatigue can be delayed by introducing a number of maneuvers which involve different elements and objectives.

Fatigue is the primary consideration in determining the length and frequency of flight instruction periods. The amount of training which can be absorbed by one student without incurring fatigue does not necessarily indicate the capacity of another student. Fatigue which results from training operations may be either physical or mental, or both. It is not necessarily a function of physical robustness or mental acuity. Generally speaking, complex operations tend to induce fatigue more rapidly than do simpler procedures regardless of the physical effort involved. Flight instruction should be continued only so long as the student is alert, receptive to instruction, and is performing at a level consistent with experience.

Students quickly become apathetic when they recognize that the instructor has made inadequate preparations for the instruction being given, or when the instruction appears to be deficient, contradictory, or insincere. To hold the student's interest and to maintain the motivation necessary for efficient learning, well-planned, appropriate, and accurate instruction must be provided. Nothing destroys a student's interest so quickly as an unplanned period of instruction. Even an inexperienced student realizes immediately when the instructor has failed to prepare a lesson. Poor preparation leads to spotty coverage, misplaced emphasis, repetition, and a complete lack of confidence on the part of the student.

THE INSTRUCTOR SHOULD ALWAYS HAVE A PLAN

Instructions may be overly explicit and so elementary as to fail to hold student interest, or they may be so general or complicated that they fail to evoke the interest necessary for effective learning. To be effective, the instructor must teach for the level of the student. The presentation must be adjusted to be meaningful to the person for whom it is intended. For example, instruction in the preflight inspection of an airplane should be presented quite differently for a student who is a skilled airplane mechanic from the instruction on the same operation for a student with no previous aeronautical experience. The inspection desired in each case is the same, but a presentation meaningful to one of these students would be inappropriate for the other.

Poor presentations of instruction may result not only from poor preparation, but also from distracting mannerisms, personal untidiness, or the appearance of irritation with the student. Creating the impression of "talking down" to the student is one of the surest ways for an instructor to lose the student's confidence and attention. Once this confidence is lost by the instructor, learning rate is unnecessarily retarded.

Anxiety may place additional burdens on the instructor. This frequently limits the student's preceptive ability and retards the development of insights.

The student must be comfortable, confident in the instructor and the airplane, and at ease, if effective learning is to occur. Providing this atmosphere for learning is one of the first and most important tasks of the instructor. Although doing so may be difficult at first, successive accomplishments of recognizable goals and the avoidance of alarming occurrences or situations will rapidly ease the student's mind. This is true of all flight students but special handling by the instructor may be required only for obvious cases.

CHAPTER XI. PLANNING INSTRUCTIONAL ACTIVITY

Any instructional activity must be competently planned and organized if it is to proceed in an effective manner. Much of the basic planning necessary for the flight and ground instructor is provided by the knowledge and proficiency requirements of the Federal Aviation Regulations, approved school syllabi, and the various texts, manuals, and training courses available. This chapter reviews briefly the planning required of the professional flight or ground instructor as it relates to three topics: (1) *course of training*, (2) *training syllabus*, and (3) *lesson plan*.

COURSE OF TRAINING

Determination of Standards and Objectives

Before any important instruction can begin, a determination of standards and objectives is necessary. In the case of a pilot training course, the overall objective is obvious, and the minimum standards are provided by Federal Aviation Regulations and flight test guides.

The general overall objective of any pilot training course is to qualify the student to be a competent, efficient, safe pilot for the operation of specific aircraft types under stated conditions. The criteria by which we determine whether the training has been adequate are the passing of written and flight tests required by the Federal Aviation Regulations for the issuance of pilot certificates.

Conscientious instructors, however, do not limit their objectives to meeting the minimum published requirements for a pilot certificate. They establish as their objectives the training of each student to have the knowledge necessary to service an airplane properly, to maneuver and operate it accurately within its limitations, and to analyze and make prompt decisions with respect to its safe operation. This is only a partial list of general objectives, but is illustrative of the major planning which is the basis of any training endeavor.

Identification of Blocks of Learning

It is not practicable for instructors to proceed immediately toward the overall objectives they have established for a major training activity being undertaken. Training for any such complicated and involved skill as piloting an aircraft requires the development and assembly, in their proper relationships, of many segments or "blocks of learning." In this way, a student can master the segments of the overall pilot performance requirements individually and can progressively combine these with other related segments until their sum meets the final objective.

Considered from this standpoint, training is much like building a pyramid—each block of learning is an identity in itself, but the pyramid is incomplete if any one block is missing. The instructor and the student must both recognize the interrelationship of the blocks and the place of each in the total objective.

After the overall training objectives have been established, the next step is the identification of the blocks of learning which constitute the necessary parts of the total objective. Just as in building a pyramid, some blocks are submerged in the structure and never appear on the surface, but each is an integral necessary part of the structure. While identifying the blocks of learning to be assembled during the proposed training activity, the planner must examine each carefully to see that it is truly an integral part of the structure. Extraneous blocks of instruction are expensive frills, especially in flight instruction, and detract from, rather than assist in, the completion of the final objective.

The blocks of learning identified during the planning of a training activity should be progressively smaller in scope. They should represent units of learning which can be measured and evaluated—not a sequence of periods of instruction. For example, the flight training of a private pilot might be divided into the following major blocks: achievement of the skills necessary for solo, the skills necessary for solo cross-country flight, and the skills appropriate for application for a private pilot certificate. Each of these, in turn, should be broken into component blocks of learning.

The skills necessary for the first solo flight might be broken down as inflight maneuvering; airspeed control, including flight at minimum controllable airspeed, stalls, and descents at approach speed; maneuvering by ground references; normal and crosswind takeoffs and landings; maximum performance operations; etc. Each of these, in turn, must be subdivided to produce effective lesson plans for each period of instruction.

As seen from the illustration cited, the possibility for breaking down and categorizing training objectives is infinite. For practical planning, the test for a useful size of a minimum block of learning is whether it contains sufficient learning to: (1) provide a challenge for the student, (2) promise a reasonable return in accomplishment for the training effort necessary, and (3) provide measurable objectives.

As these blocks of learning are completed and the student's performance of each confirmed to be at an acceptable level, the related blocks will be combined to form larger segments of the total training objective. For example, acceptable performance of airspeed management, maneuvering by ground references, inflight maneuvering, and radio communications may be combined to provide the capability of flying a traffic pattern at an airport with a control tower. In this manner, the use of a properly planned training syllabus makes it possible for the instructor to direct each period of instruction toward the completion of blocks of learning, which are in turn combined with others to lead toward the overall objective.

TRAINING SYLLABUS

The form of the syllabus may vary, but it is always in the form of an abstract or digest of the course of training. It consists of the blocks of learning to be completed in the most efficient order.

The instructor may develop a training syllabus; however, there are available many tried and proven syllabi which may be used. These are found in various training manuals, approved school syllabi, and in publications available from industry.

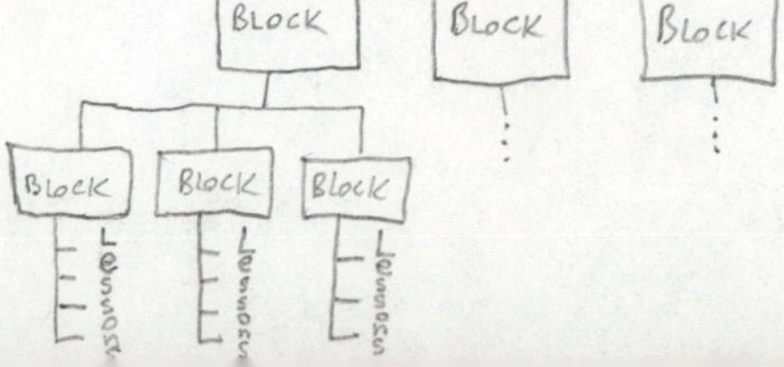

Each approved training course conducted by a certificated pilot school is given in strict accordance with a training syllabus specifically approved by the Federal Aviation Administration. Compliance with the appropriate approved syllabus is a condition for graduation from such courses. A student who has not been trained in accordance with the pertinent syllabus is not eligible for certification as an approved school graudate.

Any practical training syllabus must be flexible, and should be used primarily as a guide. The order of training can and should be altered, when necessary, to suit the progress of the student and the demands of special circumstances. In departing from the order prescribed by the syllabus, however, it is the responsibility of the instructor to consider the relationships of the blocks of learning affected. It is often preferable to skip to a completely different part of the syllabus when the conduct of a scheduled lesson is impossible, rather than proceeding to the next block, which may be predicated completely on skills to be developed during the lesson which is being postponed.

Sample Private Pilot (Airplane) Ground Training Syllabus

Each lesson of the sample private pilot (airplane) ground training syllabus which follows sets forth a unit of ground school instruction. Neither the time nor the number of ground school periods to be devoted to each lesson is specified. The sequence in which the sample lessons are listed is not necessarily the most desirable one to use in all training situations and may be varied as desired. Each lesson includes an *objective*, *content*, and *completion standards*.

Each "Lesson" is a minimum block of learning here.

LESSON NO. 1

OBJECTIVE. To develop the student's knowledge with regard to the definitions and abbreviations in Part 1 and the appropriate regulatory requirements of Part 61 of the Federal Aviation Regulations.

CONTENT.

1. Airplane Registration and Airworthiness Certificates.
2. FAR, Part 1—Definitions and abbreviations important to a private pilot.
3. FAR, Part 61.
   a. Requirements for certificates and ratings.
   b. Duration of pilot certificates.
   c. Medical certificate requirements.
   d. Written tests.
   e. Flight tests.
   f. Pilot logbooks.
   g. Recency of experience (including biennial flight review).
   h. Private pilot privileges and limitations.

COMPLETION STANDARDS. The lesson will have been successfully completed when, by an oral test, the student displays a working knowledge of the appropriate portions of FAR Part 1 and Part 61, and demonstrates the ability to locate and use information in these rules.

LESSON NO. 2

OBJECTIVE. To develop the student's knowledge of the pertinent regulatory requirements of Part 91 of the Federal Aviation Regulations and the accident reporting rules of the National Transportation Safety Board as they relate to private pilot operations.

CONTENT.

1. FAR, Part 91.
   a. General operating and flight rules.
   b. VFR requirements.
   c. IFR requirements (familiarization).
   d. Maintenance, preventive maintenance, and alterations.
   e. Familiarization with Subpart D.
2. National Transportation Safety Board Procedural Regulations, Part 830—Notification and Reporting of Accidents.

COMPLETION STANDARDS. The lesson will have been successfully completed when, by an oral test, the student demonstrates the ability to locate and use information in the appropriate rule as related to private pilot operations.

introduce, familiarize, develop, improve, review
OBJECTIVE: To develop knowledge, competence, ability, skill, understanding, confidence, proficiency
CONTENT: List
COMPLETION STANDARD: Completed when student displays working knowledge or skill, or perform, maintain V, H, ψ within... by means of test or demonstration

LESSON NO. 3

OBJECTIVE. To develop the student's knowledge of the Airman's Information Manual as it relates to VFR operations and to develop competence in using the Advisory Circular System.

CONTENT.

1. Airman's Information Manual as it relates to:
   a. Air Navigation Radio Aids.
   b. Airports and Air Navigation Lighting and Marking aids.
   c. Airspace.
   d. Air Traffic Control.
   e. Services Available to Pilots.
   f. Airport Operations.
   g. Emergency Procedures.
   h. Good Operating Practices.
   i. Airport Directory (legend).
   j. Airport Facility Directory (legend).
   k. Graphic Notices and Supplemental Data.
2. FAA Advisory Circular System—Series 00, 20, 60, 70, 90, 150, and 170 (familiarization).

COMPLETION STANDARDS. The lesson will have been successfully completed when, by an oral test and demonstration, the student displays a basic knowledge of appropriate Parts of the Airman's Information Manual and the FAA Advisory Circular System.

LESSON NO. 4

OBJECTIVE. To develop the student's knowledge of the operation of aircraft radios, the use of proper radio phraseology with respect to air traffic control facilities, and to develop competence in the use of the slide rule face of the flight computer and aeronautical charts in planning a VFR cross-country flight.

CONTENT.

1. Radio communications.
   a. Operation of radio communications equipment.
   b. Ground control.
   c. Tower.
   d. ATIS.
   e. Flight service station.
   f. UNICOM.
   g. Technique and phraseology.
2. ATC light signals.
3. Flight computer—slide rule face.
   a. Time.
   b. Speed.
   c. Distance.
   d. Fuel consumption.
4. VFR navigation.
   a. Aeronautical charts.
   b. Measurement of courses.
   c. Pilotage.
   d. Dead reckoning.

COMPLETION STANDARDS. The lesson will have been successfully completed when, by an oral test and demonstration, the student displays a basic knowledge of radio communications, ATC facilities, and aeronautical charts, and is able to use the flight computer to solve elementary VFR navigation problems.

LESSON NO. 5

OBJECTIVE. To further develop the student's knowledge of pilotage, dead reckoning, and radio navigation.

CONTENT.

1. VFR navigation.
   a. Pilotage.
   b. Dead reckoning.
2. Operation of the navigational radio equipment.
   a. VOR.
   b. ADF.
   c. Use of radio aids.
3. Flight computer—wind face.
   a. Determination of wind correction angle and true heading.
   b. Determination of groundspeed.
4. Flight computer—slide rule face.
   a. Review time, speed, and distance problems.
   b. Review fuel consumption problems.

COMPLETION STANDARDS. The lesson will have been successfully completed when, by an oral test and demonstration, the student displays a basic knowledge of VFR navigation and the use of radio aids. The student should be able to solve fundamental and advanced problems on the flight computer.

LESSON NO. 6

OBJECTIVE. To review Lesson 5 and thereby improve the student's competence in VFR navigation procedures; to introduce advanced VFR radio navigational problems; to develop the student's knowledge of emergency procedures with respect to VFR cross-country flying; and to introduce flight planning.

CONTENT.

1. Review of Lesson 5.
2. Use of ADF.
3. Radar.
4. Use of VOR, intercepting and maintaining radials.
5. Emergency procedures.
   a. Diversion to an alternate.
   b. Lost procedures, including, the use of radar and DF instructions.
   c. Inflight emergencies, including emergency landings.
6. Transponder.
7. DME.
8. Flight planning.

COMPLETION STANDARDS. The lesson will have been successfully completed when, by an oral test and demonstration, the student displays a working knowledge of advanced VFR radio navigational procedures, cross-country emergency procedures, and can accurately plan and plot a VFR cross-country flight.

LESSON NO. 7

OBJECTIVE. To further develop the student's competence in flight planning and to acquaint the student with the medical factors related to flight and general safety precautions.

CONTENT.

1. Flight planning.
2. Medical factors related to flight.
   a. Fatigue.
   b. Hypoxia.
   c. Hyperventilation.
   d. Alcohol.
   e. Drugs.
   f. Vertigo.
   g. Carbon monoxide.
3. General safety.
   a. Ground handling of aircraft.
   b. Fire—on the ground and in the air.
   c. Collision avoidance precautions.
   d. Wake turbulence avoidance.

COMPLETION STANDARDS. The lesson will have been successfully completed when, by an oral test, the student displays a basic knowledge of flight planning, the medical factors related to flight, and general safety procedures.

LESSON NO. 8

OBJECTIVE. To develop the student's knowledge of the fundamentals of weather, as associated with the operation of aircraft.

CONTENT.

1. Atmospheric layers.
2. Pressure.
3. Circulation.
4. Temperature and moisture.
5. Stability and lapse rate.
6. Turbulence.
7. Clouds.
8. Airmasses.
9. Fronts.
10. Aircraft icing.
11. Thunderstorms.

COMPLETION STANDARDS. The lesson will have been successfully completed when, by an oral test, the student demonstrates a fundamental knowledge of aviation weather.

LESSON NO. 9

OBJECTIVE. To develop the student's ability to interpret and use weather charts, reports, forecasts, and broadcasts; and to develop the student's knowledge of the procedure for obtaining weather briefings.

CONTENT.

1. Review Lesson 8.
2. Weather charts.
   a. Weather depiction charts.
   b. Surface prognostic charts.
3. Aviation weather reports.
   a. Hourly sequence reports.
   b. Special surface reports.
   c. Pilot reports.
   d. Radar reports.
4. Aviation weather broadcasts.
   a. Transcribed weather broadcasts.
   b. Inflight weather advisories.
5. Weather briefings.
6. Review requirements of regulations for VFR flight.
7. Aviation weather forecasts.
   a. Area forecasts.
   b. Terminal forecasts.
   c. Winds aloft forecasts and reports.
   d. Route forecasts.

COMPLETION STANDARDS. The lesson will have been successfully completed when, by an oral test and demonstration, the student displays the ability to interpret and use weather charts, reports, forecasts, and broadcasts, and can obtain and understand a weather briefing.

LESSON NO. 10

OBJECTIVE. To further develop the student's knowledge of aviation weather through a review of Lessons 8 and 9; to develop the student's knowledge of Greenwich time; and to develop the student's ability to recognize various weather conditions.

CONTENT.

1. Review of Lessons 8 and 9.
2. Greenwich time.
3. Weather recognition.

COMPLETION STANDARDS. The lesson will have been successfully completed when, by an oral test, the student displays a working knowledge of Greenwich time, and a knowledge of how critical weather situations can be recognized both from the ground and during flight.

LESSON NO. 11

OBJECTIVE. To develop the student's knowledge of airplane structures, propellers, engines, systems, and the magnetic compass.

CONTENT.

1. Airplane structures.
   a. Construction features.
   b. Flight control systems.
   c. Rigging.
2. Propellers.
   a. Fixed pitch.
   b. Controllable.
3. Reciprocating airplane engines.
   a. Construction features.
   b. Principle of operation—four stroke cycle.
   c. Fuel system, including carburetors and fuel injectors.
   d. Lubrication system.
   e. Ignition system.
   f. Engine instruments.
   g. Operating limitations.
   h. Malfunctions and remedial actions.
4. Airplane hydraulic system.
   a. Principle of hydraulics.
   b. Use of hydraulics in airplanes.
   c. Construction features of a simple airplane hydraulic system.
   d. Retractable landing gear and flaps.
   e. Malfunctions and remedial actions.
5. Airplane electrial system.
   a. Fundamentals of electricity.
   b. Operation of airplane electrical power system units.
   c. Electrically operated flight instruments.
   d. Retractable landing gear.
   e. Flaps.
   f. Fuses and circuit breakers.
   g. Malfunctions and remedial actions.
6. Pitot-static system and instruments.
   a. Airspeed indicator, including markings.
   b. Altimeter.
   c. Vertical-speed indicator.

7. Vacuum system and instruments.
   a. Attitude indicator.
   b. Heading indicator.
   c. Turn and slip indicator.
8. Magnetic compass.
   a. Errors.
   b. Use in flight.

COMPLETION STANDARDS. The lesson will have been successfully completed when, by an oral test, the student displays a basic understanding of airplane structures, engines, systems, and instruments.

LESSON NO. 12

OBJECTIVE. To develop the student's knowledge of basic aerodynamics.

CONTENT.

1. Forces acting on an airplane in flight.
   a. Lift.
   b. Weight.
   c. Thrust.
   d. Drag.
2. Airfoils.
   a. Angle of incidence.
   b. Angle of attack.
   c. Bernoulli's Principle.
   d. Newton's Laws.
3. Factors affecting lift and drag.
   a. Wing area.
   b. Airfoil shape.
   c. Angle of attack.
   d. Airspeed.
   e. Air density.
4. Function of the controls.
   a. Axes of rotation—longitudinal, lateral, and vertical.
   b. Primary controls—ailerons, elevators, and rudder.
   c. Secondary controls—trim tabs.
   d. Flaps and other high-lift devices.
5. Stability.
   a. Static stability.
   b. Dynamic stability.
6. Loads and load factors.
   a. Effect of bank angle on stall speed.
   b. Effect of turbulence on load factor.
   c. Effect of speed on load factor.
   d. Effect of load factor on stall speed.
7. Torque.
   a. Gyroscopic reaction.
   b. Asymmetrical loading of propeller ("P" factor).
   c. Slipstream rotation.
   d. Torque reaction.

COMPLETION STANDARDS. The lesson will have been successfully completed when, by an oral test, the student displays an understanding of basic aerodynamics.

LESSON NO. 13

OBJECTIVE. To develop the student's knowledge of the fundamental flight maneuvers.

CONTENT.

1. Straight-and-level flight.
   a. Pitch, bank, and yaw.
   b. Trim.
   c. Integrated use of outside references and flight instruments.
2. Level turns.
   a. Forces acting in a turn.
   b. Aileron drag and coordination.
   c. Speed of roll.
   d. Slips and skids.
   e. Integrated use of outside references and flight instruments.
3. Climbs and climbing turns.
   a. Best rate-of-climb airspeed.
   b. Best angle-of-climb airspeed.
   c. Torque and coordination.
   d. Trim.
4. Glides and gliding turns.
   a. Effect of high lift devices.
   b. Most efficient glide speed.
   c. Coordination.
   d. Trim.
5. Descents with power.
   a. Power settings and airspeeds.
   b. Trim.

COMPLETION STANDARDS. The lesson will have been successfully completed when, by an oral test, the student displays a basic understanding of the fundamental flight maneuvers.

LESSON NO. 14

OBJECTIVE. To develop the student's ability to properly use Pilot's Operating Handbooks and FAA Approved Airplane Flight Manuals; to develop the student's ability to perform basic weight and balance computations; and to develop the student's understanding of fundamental flight training maneuvers.

CONTENT.

1. Use of data in Pilot's Operating Handbook or FAA Approved Airplane Flight Manual.
   a. Takeoff and landing distances.
   b. Fuel consumption and related charts.
   c. Maximum range power settings.
   d. Maximum endurance power settings.
2. Weight and balance.
   a. Terms and definitions.
   b. Effects of adverse balance.
   c. Finding loaded weight.
   d. Finding center of gravity—when weight is added or removed—when weight is shifted.
3. Maneuvering at minimum controllable airspeed.
4. Stalls.
   a. Theory of stalls.
   b. Imminent stalls—power-on and power-off.
   c. Full stalls—power-on and power-off.
5. Flight maneuvering by reference to ground objects.
   a. "S" turns across a road.
   b. Rectangular course.
   c. Eights along a road.
   d. Eights across a road.
   e. Turns around a point.
   f. Eights around pylons.

COMPLETION STANDARDS. The lesson will have been successfully completed when, by an oral test and demonstration, the student displays a basic knowledge of Pilot's Operating Handbooks and FAA Approved Airplane Flight Manuals; when the student is able to perform basic weight and balance computations; and when the student has a working knowledge of the performance of fundamental flight training maneuvers.

LESSON NO. 15

OBJECTIVE. To develop the student's knowledge of fundamental flight maneuvers and attitude instrument flying.

CONTENT.

1. Review Lesson 14.
2. Takeoffs and landings.
   a. Normal and crosswind takeoffs and landings.
   b. Soft field takeoffs and landings.
   c. Short field takeoffs and landings.
   d. Go-arounds or rejected landings.
3. Introduction to attitude instrument flying. Maneuvering by reference to flight instruments—pitch, power, bank, and trim control in the performance of basic flight maneuvers.
   a. Straight-and-level flight.
   b. Turns.
   c. Climbs.
   d. Descents.
   e. Recovery from unusual attitudes.

COMPLETION STANDARDS. The lesson will have been successfully completed when, by an oral test, the student displays a basic knowledge of the performance of takeoffs and landings under various conditions, and an understanding of the performance of basic maneuvers by reference to flight instruments.

LESSON NO. 16

OBJECTIVE. To develop the student's knowledge of the fundamentals of night flying.

CONTENT.

1. Night flying—general.
   a. Requirements of regulations.
   b. Preparation.
   c. Equipment.
   d. Night vision.
   e. Airport lighting.
   f. Orientation.
   g. VFR navigation.
   h. Weather factors.
2. Partial or complete power failure.
   a. Sample situations.
   b. Recommended courses of action.

3. Systems and equipment malfunctions.

   a. Sample situations.

   b. Recommended courses of action.

COMPLETION STANDARDS. The lesson will have been successfully completed when, by an oral test, the student displays a working knowledge of the fundamentals of night flying.

Sample Private Pilot (Airplane) Flight Training Syllabus

The sample private pilot (airplane) flight training syllabus which appears on the following pages is illustrative of content and organization. It is not necessarily the most desirable syllabus to use in all training situations; however, instruction in the procedures and maneuvers listed here are considered to be most effective in the development of competence in the *pilot operations* required on the private pilot (airplane) flight test.

It should be noted that each lesson prescribes a unit of flight training, not a specified period of instruction or flight time. Each lesson also includes an *objective*, *content*, and *completion standards*. The student must have at least the instruction required by the Federal Aviation Regulations before the first solo and the first solo cross-country flights, and the flight experience prescribed for a private pilot certificate at the completion of the syllabus.

The notation "(VR and IR)" is used to indicate maneuvers which should be performed by both visual references and instrument references during the conduct of integrated flight instruction.

Throughout the student's flight training, the instructor should emphasize collision and wake turbulence avoidance procedures.

LESSON NO. 1—DUAL

OBJECTIVE. To familiarize the student with the training airplane, its servicing, its operating characteristics, cabin controls, instruments, systems, preflight procedures, use of checklists, and safety precautions to be followed; to acquaint the student with the sensations of flight and the effect and use of controls; and to familiarize the student with the local flying area and airport.

CONTENT.

1. Preflight discussion.

2. Introduction.

   a. Airplane servicing.

   b. Purpose of preflight checks.

   c. Visual inspection.

   d. Importance of using a checklist.

   e. Engine starting procedure.

   f. Radio communications procedures.

   g. Taxiing.

   h. Pretakeoff checklist.

   i. Takeoff.

   j. Traffic pattern departure, climb-out, and level-off.

   k. Effect and use of controls (VR and IR).

   l. Straight-and-level flight (VR and IR).

   m. Medium bank turns (VR and IR).

   n. Local flying area familiarization.

   o. Collision avoidance.

   p. Wake turbulence avoidance.

   q. Traffic pattern entry, approach, landing, and parking.

   r. Ground safety.

3. Postflight critique and preview of next lesson.

COMPLETION STANDARDS. The lesson will have been successfully completed when the student understands how to service the airplane, the use of a checklist for the visual inspection, starting procedure, and engine run-up; displays a knowledge of the effect and use of controls; and has a reasonable familiarity with the local flying area and airport.

LESSON NO. 2—DUAL

OBJECTIVE. To develop the student's skill in the performance of the four basic flight maneuvers (climbs, descents, turns, and straight-and-level flight).

CONTENT.

1. Preflight discussion.

2. Review.

   a. Airplane servicing.

   b. Visual inspection.

   c. Engine starting procedure.

   d. Radio communications procedures.

e. Taxiing.
f. Pretakeoff checklist.
g. Takeoff.
h. Traffic pattern departure.
i. Straight-and-level flight (VR and IR).
j. Medium bank turns (VR and IR).
k. Traffic pattern entry, approach, landing, and parking.

3. Introduction.
   a. Climbs and climbing turns (VR and IR).
   b. Glides and gliding turns (VR and IR).
   c. Torque effect.
   d. Level-off from climbs and glides (VR and IR).
4. Postflight critique and preview of next lesson.

COMPLETION STANDARDS. The lesson will have been successfully completed when the student can perform, with minimum assistance from the instructor, climbs, straight-and-level flight, turns, and glides. During straight-and-level flight the student should, with minimum instructor assistance, be able to maintain altitude within ±100 feet, airspeed within ±10 knots, and heading within ±10° of that assigned.

LESSON NO. 3—DUAL

OBJECTIVE. To review lessons One and Two; to develop the student's proficiency in the performance of the basic flight maneuvers; and to introduce maneuvering at minimum controllable airspeed and power-off stalls.

CONTENT.

1. Preflight discussion.
2. Review.
   a. Use of checklist.
   b. Engine starting procedure.
   c. Radio communications procedures.
   d. Takeoff.
   e. Traffic pattern departure.
   f. Climbs and climbing turns (VR and IR).
   g. Straight-and-level flight (VR and IR).
   h. Medium bank turns (VR and IR).
   i. Glides and gliding turns (VR and IR).
   j. Level-off procedures (VR and IR).
   k. Traffic pattern and landing.
3. Introduction.
   a. Maneuvering at minimum controllable airspeed (VR and IR).
   b. Power-off stalls (imminent and full) (VR and IR).
   c. Descents and descending turns, with power (VR and IR).
4. Postflight critique and preview of next lesson.

COMPLETION STANDARDS. The lesson will have been successfully completed when the student can display reasonable proficiency in the performance of the four basic flight maneuvers, and perform with minimum assistance, flight at minimum controllable airspeed. During this and subsequent flight lessons, the student should be able to perform the visual inspection, starting procedure, radio communications, taxiing, pretakeoff check, parking, and shut-down procedure without assistance. During climbs, level flight, turns, glides, and maneuvering at minimum controllable airspeed the student should, with minimum instructor assistance, be able to maintain assigned airspeed within ±10 knots. The student should also, with minimum instructor assistance, be able to maintain assigned altitude within ±100 feet and assigned heading within ±10°.

LESSON NO. 4—DUAL

OBJECTIVE. To review previous lessons, thereby increasing the student's competence in the performance of fundamental flight maneuvers; and to introduce power-on stalls, rectangular course, S-turns across a road, eights along a road, and elementary emergency landings.

CONTENT.

1. Preflight discussion.
2. Review.
   a. Takeoff.
   b. Traffic pattern departure.
   c. Climbs and climbing turns (VR and IR).
   d. Straight-and-level flight and medium bank turns (VR and IR).
   e. Maneuvering at minimum controllable airspeed (VR and IR).
   f. Power-off stalls (imminent and full) (VR and IR).
   g. Glides and gliding turns (VR and IR).

h. Descents and descending turns, with power (VR and IR).

i. Level-off procedures (VR and IR).

j. Traffic pattern and landing.

3. Introduction.

a. Power-on stalls (imminent and full) (VR and IR).

b. Rectangular course.

c. S-turns across a road.

d. Eights along a road and eights across a road.

e. Elementary emergency landings.

4. Postflight critique and preview of next lesson.

COMPLETION STANDARDS. The lesson will have been successfully completed when the student is competent to perform, with minimum instructor assistance, the procedures and maneuvers given during previous lessons. The student should achieve the ability to recognize stall indications and make safe prompt recoveries. The student should maintain assigned airspeed within ±10 knots, assigned altitude within ±100 feet, and assigned heading within ±10°, and display a basic knowledge of elementary emergency landings.

LESSON NO. 5—DUAL

OBJECTIVE. To review previous lessons, with emphasis on maneuvering by reference to ground objects. To develop the student's ability to perform climbs at best rate and best angle, crosswind takeoffs and landings; and to introduce emergency procedures, changes of airspeed and configuration, turns around a point, and eights around pylons.

CONTENT.

1. Preflight discussion.

2. Review.

a. Takeoff.

b. Climbs and climbing turns (VR and IR).

c. Maneuvering at minimum controllable airspeed (VR and IR).

d. Power-off and power-on stalls (imminent and full).

e. Rectangular course.

f. S-turns across a road.

g. Eights along a road.

h. Elementary emergency landings.

i. Traffic pattern and landing.

3. Introduction.

a. Crosswind takeoffs and landings.

b. Climb at best rate (VR and IR).

c. Climb at best angle (VR and IR).

d. Emergency procedures.

e. Change of airspeed and configuration (VR and IR).

f. Turns around a point.

g. Eights around pylons.

4. Postflight critique and preview of next lesson.

COMPLETION STANDARDS. The lesson will have been successfully completed when the student can recognize imminent and full stalls and make prompt effective recoveries, perform ground reference maneuvers with reasonably accurate wind drift corrections and good coordination, and has a proper concept of crosswind technique during takeoffs and landings. The student should have a working knowledge of emergency procedures, and be able to perform them with minimum assistance. During ground reference maneuvers, the student should maintain airspeed within ±10 knots, altitude within ±100 feet, and heading within ±10° of that desired.

LESSON NO. 6—DUAL

OBJECTIVE. To review previous lessons; to develop the student's ability to perform slips, accelerated stalls, cross-control stalls, and advanced emergency landings; to improve the student's proficiency in normal and crosswind takeoffs and landings; and to introduce balked takeoffs and go-arounds (rejected landings).

CONTENT.

1. Preflight discussion.

2. Review.

a. Normal and crosswind takeoffs.

b. Climbs at best rate and best angle (VR and IR).

c. Power-off stalls (imminent and full) (VR and IR).

d. Power-on stalls (imminent and full) (VR and IR).

e. Change of airspeed and configuration (VR and IR).
f. Turns around a point.
g. Eights around pylons.
h. Emergency procedures.
i. Normal and crosswind landings.

3. Introduction.
a. Balked takeoffs.
b. Accelerated stalls.
c. Cross-control stalls.
d. 180° and 360° gliding approaches.
e. Advanced emergency landings.
f. Side slips and forward slips.
g. Go-arounds (rejected landings).

4. Postflight critique and preview of next lesson.

COMPLETION STANDARDS. The lesson will have been successfully completed when the student can perform stall recoveries smoothly and promptly with a minimum loss of altitude, is able to make unassisted normal and crosswind takeoffs and landings, and can plan and fly emergency landing patterns with accuracy and consistency. The student should be able to execute balked takeoffs and go-arounds (rejected landings) without assistance, and should maintain assigned airspeed within ±10 knots, assigned altitude within ±100 feet, and assigned heading within ±10°.

LESSON NO. 7—DUAL

OBJECTIVE. To review previous lessons. To further develop the student's competence in takeoffs, traffic patterns, and landings through concentrated practice. To develop the student's ability to use slips during landing approaches and improve the ability to perform go-arounds (rejected landings).

CONTENT.

1. Preflight discussion.

2. Review.
a. Normal and crosswind takeoffs.
b. Normal and crosswind landings (touch-and-go and full-stop).
c. Forward slips.
d. Go-arounds (rejected landings).
e. 180° and 360° gliding approaches.
f. Advanced emergency landings.
g. Emergency procedures.

3. Postflight critique and preview of next lesson.

COMPLETION STANDARDS. The lesson will have been successfully completed when the student can fly accurate traffic patterns and make unassisted normal and crosswind takeoffs and landings. The student should be competent in the go-around (rejected landing) procedure. During traffic patterns, the student should maintain desired airspeed within ±10 knots, desired altitude within ±100 feet, and desired heading within ±10°.

LESSON NO. 8—DUAL

OBJECTIVE. To review power-off stalls, maneuvering at minimum controllable airspeed, and advanced emergency landings. To continue to develop the student's competence in takeoffs, traffic patterns, and landings, and to improve the ability to recover from poor approaches and landings.

CONTENT

1. Preflight discussion.

2. Review.
a. Normal and crosswind takeoffs.
b. Power-off stalls (imminent and full) (VR and IR).
c. Maneuvering at minimum controllable airspeed (VR and IR).
d. Advanced emergency landings.
e. Normal and crosswind landings (touch-and-go and full-stop).
f. Go-arounds (rejected landings).
g. Recovery from poor approaches and landings.

3. Postflight critique and preview of next lesson.

COMPLETION STANDARDS. The lesson will have been successfully completed when the student can demonstrate a degree of proficiency in normal and crosswind takeoffs and landings and traffic patterns, which is considered safe for solo. The student should display sound judgment and proper techniques in recoveries from poor approaches and landings. During traffic patterns, the student should maintain desired airspeed within ±10 knots, desired altitude within ±100 feet, and desired heading within ±10°.

LESSON NO. 9—DUAL AND SOLO

OBJECTIVE. To develop the student's competence to a level which will allow the safe accomplishment of the first supervised solo in the traffic pattern.

CONTENT.

1. Preflight discussion.
2. Review.
   a. Normal and crosswind takeoffs.
   b. Normal and crosswind landings (full-stop).
   c. Go-arounds (rejected landings).
   d. Recovery from poor approaches and landings.
   e. Elementary emergency landings.
3. Introduction—first supervised solo in the traffic pattern. Three takeoffs and three full-stop landings should be performed.
4. Postflight critique and preview of next lesson.

COMPLETION STANDARDS. The lesson will have been successfully completed when the student safely accomplishes the first supervised solo in the traffic pattern.

LESSON NO. 10—DUAL AND SOLO

OBJECTIVE. To review previous lessons and to accomplish the student's second supervised solo in the traffic pattern.

CONTENT.

1. Preflight discussion.
2. Review.
   a. Takeoff and traffic departure.
   b. Climbs and climbing turns (VR and IR).
   c. Maneuvering at minimum controllable airspeed (VR and IR).
   d. Power-off stalls (imminent and full) (VR and IR).
   e. Advanced emergency landings.
   f. Traffic patterns, approaches and landings.
   g. Recovery from poor approaches and landings.
3. Introduction—second supervised solo in the traffic pattern. Three takeoffs, two touch-and-go, and one full-stop landing should be performed.
4. Postflight critique and preview of next lesson.

COMPLETION STANDARDS. The lesson will have been successfully completed when the student demonstrates solo competence in maneuvers performed and safely accomplishes the second supervised solo in the traffic pattern.

LESSON NO. 11—DUAL AND SOLO

OBJECTIVE. To review presolo maneuvers with higher levels of proficiency required. To introduce short and soft field takeoffs, and maximum climbs; and to accomplish the student's third supervised solo in the traffic pattern.

CONTENT.

1. Preflight discussion.
2. Review.
   a. Selected presolo maneuvers (VR and IR).
   b. Takeoffs, traffic patterns, and landings.
   c. Balked takeoff.
   d. Go-around (rejected landing).
   e. Recovery from poor approach and landing.
3. Introduction.
   a. Short field takeoffs and maximum climbs.
   b. Soft field takeoffs.
   c. Third supervised solo in the traffic pattern. At least three takeoffs and landings should be performed.
4. Postflight critique and preview of next lesson.

COMPLETION STANDARDS. The lesson will have been successfully completed when the student demonstrates solo competence in the selected presolo maneuvers performed and safely accomplishes the third supervised solo in the traffic pattern. The student should be able to perform short field takeoffs, soft field takeoffs, and maximum climbs without instructor assistance.

LESSON NO. 12—DUAL

OBJECTIVE. To refamiliarize the student with the local practice area and to improve proficiency in the presolo maneuvers in preparation for local area solo practice flights. To develop the student's ability to obtain radar and DF heading instructions and to become oriented in relation to a VOR, and to "home" to a nondirectional beacon using ADF. To introduce wheel landings (tail wheel airplanes).

CONTENT.

1. Preflight discussion.
2. Review.
   a. Practice area orientation.
   b. Power-off stalls (imminent and full) (VR and IR).
   c. Power-on stalls (imminent and full) (VR and IR).
   d. Maneuvering at minimum controllable airspeed (VR and IR).
   e. Turns around a point.
   f. Eights around pylons.
   g. Crosswind takeoffs and landings.
   h. 180° and 360° gliding approaches.
   i. Advanced emergency landings.
   j. Emergency procedures.
3. Introduction.
   a. Use of radar and DF heading instructions (VR and IR).
   b. VOR orientation (VR and IR).
   c. ADF "homing" (VR and IR).
   d. Wheel landings (tailwheel airplanes).
4. Postflight critique and preview of next lesson.

COMPLETION STANDARDS. The lesson will have been successfully completed when the student demonstrates an improved performance of the presolo maneuvers, is able to determine position in the local practice area by pilotage, VOR, or ADF; and can safely perform assigned maneuvers. The student should be competent in obtaining radar and DF heading instructions and in the performance of simulated emergency landings and emergency procedures.

LESSON NO. 13—SOLO

OBJECTIVE. To develop the student's confidence and proficiency through solo practice of assigned maneuvers.

CONTENT.

1. Preflight discussion.
2. Review.
   a. Normal and/or crosswind takeoffs and landings.
   b. Power-off stalls (imminent and full).
   c. Power-on stalls (imminent and full).
   d. Maneuvering at minimum controllable airspeed.
   e. Other maneuvers specified by the instructor during the preflight discussion.
3. Postflight critique and preview of next lesson.

COMPLETION STANDARDS. The lesson will have been successfully completed when the student has accomplished the solo review and practiced the basic and precision flight maneuvers, in addition to those specified by the instructor. The student should gain confidence and improve flying technique as a result of the solo practice period.

LESSON NO. 14—DUAL

OBJECTIVE. To improve the student's proficiency in previously covered procedures and maneuvers and to review advanced emergency landings, emergency procedures, and orientation by means of VOR and/or ADF.

CONTENT.

1. Preflight discussion.
2. Review.
   a. Normal and/or crosswind takeoffs and landings.
   b. Power-off stalls (imminent and full) (VR and IR).
   c. Power-on stalls (imminent and full) (VR and IR).
   d. Maneuvering at minimum controllable airspeed (VR and IR).
   e. Accelerated stalls.
   f. Eights around pylons.
   g. Short field and soft field takeoffs and landings.
   h. Advanced emergency landings.
   i. Emergency procedures.
   j. Orientation by means of VOR and/or ADF.

3. Postflight critique and preview of next lesson.

COMPLETION STANDARDS. The lesson will have been successfully completed when the student demonstrates an increased proficiency in previously covered procedures and maneuvers. The student should be able to maintain airspeed within ±10 knots, altitude within ±100 feet, and heading within ±10° of that desired.

LESSON NO. 15—SOLO

OBJECTIVE. To further develop the student's confidence and proficiency through solo practice of assigned maneuvers.

CONTENT.

1. Preflight discussion.
2. Review.
   a. Normal and/or crosswind takeoffs and landings.
   b. Turns around a point.
   c. Eights around pylons.
   d. Other maneuvers specified by the instructor during the preflight discussion.
3. Postflight critique and preview of next lesson.

COMPLETION STANDARDS. The lesson will have been successfully completed when the student has accomplished the solo review and thereby increased proficiency and confidence.

LESSON NO. 16—DUAL

OBJECTIVE. To develop the student's ability to plan, plot, and fly a 2-hour day cross-country flight with landings at two unfamiliar airports; to develop the student's proficiency in navigating by means of pilotage, dead reckoning, VOR, and/or ADF; and to develop the ability to take proper action in emergency situations.

CONTENT.

1. Preflight discussion.
   a. Planning flight, including weather check.
   b. Plotting course.
   c. Preparing log.
   d. Filing and closing VFR flight plan.
2. Introduction.
   a. Filing VFR flight plan.
   b. Pilotage.
   c. Dead reckoning.
   d. Tracking VOR radial and/or homing by ADF (VR and IR).
   e. Departure, en route, and arrival radio communications.
   f. Simulated diversion to an alternate airport.
   g. Unfamiliar airport procedures.
   h. Emergencies, including DF and radar heading instructions (VR and IR).
   i. Closing VFR flight plan.
3. Postflight critique and preview of next lesson.

COMPLETION STANDARDS. The lesson will have been successfully completed when, with instructor assistance, the student is able to perform the cross-country preflight planning, fly the planned course making necessary off-course corrections, and can make appropriate radio communications. The student should be competent in navigating by means of pilotage, dead reckoning, VOR, and/or ADF, and when so instructed, is able to accurately plan and fly a diversion to an alternate airport.

LESSON NO. 17—DUAL

OBJECTIVE. To improve the student's proficiency in cross-country operations through the planning, plotting, and flying of a second dual 2-hour day cross-country flight, with landings at two unfamiliar airports. To improve the student's competence in navigating by means of pilotage, dead reckoning, VOR, and ADF; and to further develop the ability to take proper action in emergency situations.

CONTENT.

1. Preflight discussion.
   a. Planning flight, including weather check.
   b. Plotting course.
   c. Preparing log.
   d. Filing and closing VFR flight plan.
2. Review.
   a. Filing VFR flight plan.
   b. Pilotage and dead reckoning.

c. Radio navigation (VOR and/or ADF) (VR and IR).
d. Departure, en route, and arrival radio communications.
e. Simulated diversion to an alternate airport.
f. Unfamiliar airport procedures.
g. Emergencies, including DF and radar heading instructions (VR and IR).
h. Closing VFR flight plan.

3. Postflight critique and preview of next lesson.

COMPLETION STANDARDS. The lesson will have been successfully completed when the student, with minimum instructor assistance, is able to plan, plot, and fly the planned course. Estimated times of arrival should be accurate with an apparent error of not more than 10 minutes. Any off-course corrections should be accomplished accurately and promptly. The student should be able to give the instructor an accurate position report at any time without hesitation. When given a "simulated lost" situation, the student should be able to initiate and follow an apppropriate "lost procedure."

LESSON NO. 18—SOLO

OBJECTIVE. To develop the student's ability to plan, plot, and fly a 3-hour solo day cross-country flight, with landings at two unfamiliar airports, thereby improving proficiency and confidence in the conduct of future solo cross-country flights. To improve the student's proficiency in navigating by means of pilotage, dead reckoning, VOR, and/or ADF; and to increase the ability to cope with new or unexpected flight situations.

CONTENT.

1. Preflight discussion.
a. Planning flight, including weather check.
b. Plotting course.
c. Preparing log.
d. Filing and closing VFR flight plan.
e. Procedure at unfamiliar airports.
f. Emergencies.

2. Review.
a. Filing VFR flight plan.
b. Pilotage.
c. Dead reckoning.
d. Radio navigation (VOR and/or ADF).
e. Departure, en route, and arrival radio communications.
f. Unfamiliar airport procedures.
g. Closing VFR flight plan.

3. Postflight critique and preview of next lesson.

COMPLETION STANDARDS. The lesson will have been successfully completed when the student is able to plan, plot, and fly the 3-hour cross-country flight as assigned by the instructor. The instructor should determine how well the flight was conducted through oral questioning.

LESSON NO. 19—DUAL AND SOLO

OBJECTIVE. To develop the student's ability to make solo night flights in the local practice area and airport traffic pattern. To familiarize the student with such aspects of night operations as: night vision, night orientation, judgment of distance, use of cockpit lights, position lights, landing lights, and night emergency procedures.

CONTENT.

1. Preflight discussion.
a. Night vision and vertigo.
b. Orientation in local area.
c. Judgment of distance.
d. Aircraft lights.
e. Airport lights.
f. Taxi technique.
g. Takeoff and landing technique.
h. Collision avoidance.
i. Unusual attitude recovery.
j. Emergencies.

2. Introduction.
a. Night visual inspection.
b. Use of cockpit lights.
c. Taxi techniques.
d. Takeoff and traffic departure.
e. Area orientation.
f. Interpretation of aircraft and airport lights.

g. Recovery from unusual attitudes (VR and IR).

h. Radio communications.

i. Traffic entry.

j. Power approaches and full-stop landings.

k. Use of landing lights.

l. Simulated electrical failure to include at least one black-out landing.

3. Postflight critique and preview of next lesson.

COMPLETION STANDARDS. The lesson will have been successfully completed when the student displays the ability to maintain orientation in the local flying area and traffic pattern, can accurately interpret aircraft and runway lights, and can competently fly the traffic pattern and perform takeoffs and landings. The student should display, through oral quizzing and demonstrations, competence in performing night emergency procedures. At least five takeoffs and landings should be accomplished.

LESSON NO. 20—DUAL

OBJECTIVE. To develop the student's ability to plan, plot, and fly a $1\frac{1}{2}$-hour night cross-country flight around a triangular course with at least one landing at an unfamiliar airport. To develop the student's competence in navigating at night by means of pilotage, dead reckoning, and VOR or ADF; and to develop the student's ability to take proper action in night emergency situations.

CONTENT.

1. Preflight discussion.
   a. Planning $1\frac{1}{2}$-hour night cross-country flight, including weather check.
   b. Plotting course.
   c. Preparing log.
   d. Filing and closing VFR flight plan.

2. Introduction.
   a. Filing VFR flight plan.
   b. Proper use of cockpit lights and flashlight for chart reading.
   c. Pilotage—factors peculiar to night flying.
   d. Dead reckoning.
   e. Tracking VOR radial and/or homing by ADF.
   f. Departure, en route, and arrival radio communications.
   g. Simulated diversion to an alternate airport.
   h. Emergencies, including simulated failure of electrical system, also DF and radar heading instructions.
   i. Closing VFR flight plan.

3. Postflight critique and preview of next lesson.

COMPLETION STANDARDS. The lesson will have been successfully completed when, with minimum assistance from the instructor, the student is able to perform the night cross-country preflight planning, fly the planned course making necessary off-course corrections, and can make appropriate radio communications. The student should be competent in navigating by means of pilotage, dead reckoning, and VOR or ADF. The student should have a thorough knowledge of night emergency procedures.

LESSON NO. 21—SOLO

OBJECTIVE. To further develop the student's competence in cross-country operations through the planning, plotting, and flying of a second solo 3-hour day cross-country flight with landings at two unfamiliar airports. To improve the student's proficiency in navigating by means of pilotage, dead reckoning, VOR, and/or ADF; and to further increase the student's confidence and ability to properly handle unexpected flight situations.

CONTENT.

1. Preflight discussion.
   a. Planning flight, including weather check.
   b. Plotting course.
   c. Preparing log.
   d. Filing and closing VFR flight plan.
   e. Procedure at unfamiliar airports.
   f. Emergencies.

2. Review.
   a. Filing VFR flight plan.
   b. Pilotage and dead reckoning.
   c. Radio navigation (VOR and/or ADF).

d. Departure, en route, and arrival radio procedures.
e. Unfamiliar airport procedures.
f. Closing VFR flight plan.

3. Postflight critique and preview of next lesson.

COMPLETION STANDARDS. The lesson will have been successfully completed when the student is able to plan, plot, and fly the second 3-hour day cross-country flight as assigned by the instructor. The instructor should determine how well the flight was conducted through oral questioning.

LESSON NO. 22—SOLO

OBJECTIVE. To further develop the student's competence in cross-country operations through the planning, plotting, and flying of a solo 4-hour day cross-country flight, with landings at three unfamiliar airports, each of which is more than 100 nautical miles from the other airports.

CONTENT.

1. Preflight discussion.
   a. Planning flight, including weather check.
   b. Plotting course.
   c. Preparing log.
   d. Filing and closing VFR flight plan.
   e. Procedure at unfamiliar airports.
   f. Emergencies.
2. Review.
   a. Filing VFR flight plan.
   b. Pilotage and dead reckoning.
   c. Radio navigation (VOR and/or ADF).
   d. Departure, en route, and arrival radio communications.
   e. Unfamiliar airport procedures.
   f. Closing VFR flight plan.
3. Postflight critique and preview of next lesson.

COMPLETION STANDARDS. The lesson will have been successfully completed when the student is able to plan, plot, and fly the 4-hour day cross-country flight as assigned by the instructor. The instructor should determine how well the flight was conducted through oral questioning.

LESSON NO. 23—DUAL

OBJECTIVE. To develop precision in the student's performance of procedures and maneuvers covered previously with emphasis directed to stalls.

CONTENT.

1. Preflight discussion.
2. Review.
   a. Power-off stalls (imminent and full) (VR and IR).
   b. Power-on stalls (imminent and full) (VR and IR).
   c. Maneuvering at minimum controllable airspeed (VR and IR).
   d. 180° and 360° gliding approaches.
   e. Advanced emergency landings.
   f. Slips.
   g. Crosswind takeoffs and landings.
   h. Short field and soft field takeoffs and landings.
   i. Emergency procedures.
3. Introduction of ASR approaches.
4. Postflight critique and preview of next lesson.

COMPLETION STANDARDS. The lesson will have been successfully completed when the student demonstrates improved performance in the various maneuvers given. The student should be able to make ASR approaches with minimum instructor assistance.

LESSON NO. 24—SOLO

OBJECTIVE. To further develop the student's competence through solo practice of assigned maneuvers. Emphasis will be directed to stalls.

CONTENT.

1. Preflight discussion.
2. Review.
   a. Power-on and power-off stalls (imminent and full).
   b. Maneuvering at minimum controllable airspeed.
   c. Short field and soft field takeoffs and landings.

d. Other maneuvers assigned by the instructor during preflight discussion.

3. Postflight critique and preview of next lesson.

COMPLETION STANDARDS. The lesson will have been successfully completed when the student has accomplished the solo review and practiced the basic and precision flight maneuvers in addition to those specified by the instructor. The student should gain confidence and improve flying technique as a result of the solo practice period.

LESSON NO. 25—DUAL

OBJECTIVE. To develop improved performance and precision in the procedures and maneuvers covered previously with emphasis directed to ground track maneuvers.

CONTENT.

1. Preflight discussion.
2. Review.
   a. Maneuvering at minimum controllable airspeed.
   b. Turns around a point.
   c. Eights around pylons.
   d. 180° and 360° gliding approaches.
   e. Advanced emergency landings.
   f. Slips.
   g. Crosswind takeoffs and landings.
   h. Wheel landings (tail wheel airplane).
   i. ASR approach.
3. Postflight critique and preview of next lesson.

COMPLETION STANDARDS. The lesson will have been successfully completed when the student demonstrates improved performance in the maneuvers given.

LESSON NO. 26—SOLO

OBJECTIVE. To further develop the student's competence through solo practice of assigned maneuvers. Emphasis will be directed to ground track maneuvers.

CONTENT.

1. Preflight discussion.
2. Review.
   a. Turns around a point.
   b. Eights around pylons.
   c. Short and soft field takeoffs and landings.
   d. Wheel landings (tail wheel airplanes).
   e. Other maneuvers assigned by the instructor during the preflight discussion.
3. Postflight critique and preview of next lesson.

COMPLETION STANDARDS. The lesson will have been successfully completed when the student has accomplished the solo review. The student should gain proficiency in the ground track and other maneuvers assigned by the instructor.

LESSON NO. 27—SOLO

OBJECTIVE. To improve the student's proficiency in the pilot operations required on the private pilot (airplane) flight check.

CONTENT.

1. Preflight discussion.
2. Review.
   a. Ground track maneuvers.
   b. Power-on and power-off stalls (imminent and full).
   c. Maneuvering at minimum controllable airspeed.
   d. Crosswind takeoffs and landings.
   e. Other maneuvers assigned by the instructor during the preflight discussion.
3. Postflight critique and preview of next lesson.

COMPLETION STANDARDS. The lesson will have been successfully completed when the student has gained proficiency in the procedures and maneuvers assigned by the instructor.

LESSON NO. 28—DUAL

OBJECTIVE. To evaluate the student's performance of the procedures and maneuvers necessary to conduct flight operations as a private pilot.

CONTENT.

1. Preflight discussion.

2. Review.
   a. Power-on and power-off stalls (imminent and full).
   b. Maneuvering at minimum controllable airspeed.
   c. Ground track maneuvers.
   d. 180° and 360° gliding approaches.
   e. Advanced emergency landings.
   f. Short field and soft field takeoffs and landings.
   g. Crosswind takeoffs and landings.
   h. Straight-and-level flight, turns, climbs, descents, and recovery from unusual attitudes by reference to flight instruments.
   i. Tracking VOR radial and homing by ADF (VR and IR).
   j. Use of radar and DF heading instructions (VR and IR).
   k. ASR approach (VR and IR).
   l. Emergency operations.

3. Post-flight critique

COMPLETION STANDARDS. The lesson will have been successfully completed when the student satisfactorily performs the procedures and maneuvers selected to show competence in the pilot operations listed in the Private Pilot (Airplane) Flight Test Guide.

NOTE: Before signing a student's flight-test recommendation, it is the responsibility of the flight instructor to see that all the aeronautical experience requirements for a private pilot certificate are met.

LESSON PLAN

A lesson plan is an organized outline or "blueprint" for a single instructional period and should be prepared in written form for each ground school and flight period, regardless of the instructor's experience. A lesson plan should be developed to show specific knowledge and/or skills to be taught. It is a necessary guide for the instructor in that it tells *what to do*, in *what order to do it*, and *what procedure to use* in teaching the material of the lesson.

A so-called "mental outline" of a lesson is *not* a lesson plan. A lesson plan must be put into writing. Another instructor should be able to take the lesson plan and know what to do in conducting the same period of instruction. When placed in writing, the lesson plan can be analyzed from the standpoint of adequacy and completeness.

Purpose of the Lesson Plan

Lesson plans are designed to assure that each student receives the best possible instruction under the existing conditions. Lesson plans help instructors keep a constant check on their own activity, as well as that of their students. The development of lesson plans by instructors signifies, in effect, that they have taught the lessons to themselves prior to attempting to teach the lessons to students. An adequate lesson plan, when properly used, should:

1. Assure a wise selection of material and the elimination of unimportant details.
2. Make certain that due consideration is given to each part of the lesson.
3. Aid the instructor in presenting the material in a suitable sequence for efficient learning.
4. Provide an outline of the teaching procedure to be used.
5. Serve as a means of relating the lesson to the objectives of the course of training.
6. Give the inexperienced instructor confidence.
7. Promote uniformity of instruction regardless of the instructor or the date on which the lesson is given.

Characteristics of a Well-Planned Lesson

1. *Unity.* Each lesson should be a unified segment of instruction. A lesson is concerned with certain limited objectives which are stated in terms of desired student learning outcomes. All teaching procedures and materials should be selected to attain these objectives.
2. *Content.* Each lesson should contain new material. However, the new facts, principles, procedures, or skills should be related to the lesson previously presented. A short review of earlier lessons is usually necessary, particularly in flight training.
3. *Scope.* Each lesson should be reasonable in scope. A person can master only a few principles or skills at a time, the number

depending on complexity. Presenting too much material in a lesson results in confusion; presenting too little material results in inefficiency.

4. *Practicality.* Each lesson should be planned in terms of the conditions under which the training is to be conducted. Lesson plans conducted in an airplane or ground trainer will differ from those conducted in a classroom. Also, the kinds and quantities of instructional aids available have a great influence on lesson planning and instructional procedures.

5. *Relation to Course of Training.* Each lesson should be planned and taught so that its relation to the course objectives are clear to each student. For example, a lesson on short field takeoffs and landings should be related to both the certification and safety objectives of the course of training.

6. *Instructional Steps.* Every lesson, when adequately developed, falls logically into the four steps of the teaching process; i.e., preparation, presentation, application, and review and evaluation.

How to Use a Lesson Plan Properly

1. *Be Familiar With the Lesson Plan.* The instructor should study each step of the plan and should be thoroughly familiar with as much information related to the subject as possible.

2. *Use the Lesson Plan as a Guide.* The lesson plan is an outline for conducting an instructional period. It assures that pertinent materials are at hand and that the presentation is accomplished with order and unity. Having a plan prevents the instructor from "getting off the track," omitting essential points, and introducing irrelevant material. Students have a right to expect an instructor to give the same attention to teaching that they give to learning. The most certain means of achieving teaching success is to have a carefully thought-out lesson plan.

3. *The Lesson Plan is not a Substitute for Thinking.* Instructors should always know more than they have time to teach. The lesson plan is a framework or skeleton; the instructor should fill it out with as many relevant examples and practical applications as possible.

4. *Adapt the Lesson Plan to the Class or Student.* In teaching a ground school period, the instructor may find that the procedures outlined in the lesson plan are not leading to the desired results. In this situation, the instructor should change the approach. There is no certain way of predicting the reactions of different groups of students. An approach which has been successful with one group may not be equally successful with another.

   A lesson plan for an instructional flight period should be appropriate to the background, flight experience, and ability of the particular student. A rigidly prepared lesson plan should *not* be used for an instructional flight because each student requires a slightly different approach. A lesson plan may have to be modified considerably during flight, due to deficiencies in the student's knowledge or poor mastery of elements essential to the effective completion of the lesson. In some cases, the entire lesson plan may have to be abandoned in favor of review.

5. *Revise the Lesson Plan Periodically.* After a lesson plan has been prepared for a ground school period, a continuous revision will be necessary. This is true for a number of reasons; e.g., availability or nonavailability of instructional aids; changes in regulations, new manuals and textbooks; changes in the state-of-the-art; etc.

Lesson Plan Items

Any lesson plan, whether it is for a ground school period or an instructional flight, should contain the following items:

1. *Lesson objective.* The objective of the lesson should be clearly stated in terms of desired student learning outcomes. The objective is the reason for the lesson—what the instructor expects the student to know or do at the completion of the lesson.

The objective for a ground school period on "maneuvering by reference to flight instruments" could be, "To develop the student's understanding of attitude instrument flying as related to straight-and-level flight, climbs and descents, and recovery from unusual attitudes." The objective for an instructional flight period on "ground reference maneuvers" could be, "To develop the student's skill in planning and following a pattern over the ground compensating for wind drift at varying angles."

2. *Elements involved.* This is a statement of the elements of knowledge and skill necessary for the fulfillment of the lesson objective. This may include both elements previously learned and those to be introduced during this lesson. A statement of the elements of a ground school lesson on "maneuvering by reference to flight instruments" should include: (a) straight-and-level flight, (b) turns, (c) climbs and descents, and (d) recovery from unusual attitudes.

"List of Contents" of knowledge & skill cf. "Content" of each lesson in the training syllabus

The elements of an instructional flight period on "ground reference maneuvers" could be: (a) use of ground references to control path, (b) observation and control of wind effect, and (c) control of airplane attitude, altitude, and heading.

3. *Schedule.* The instructor should estimate the amount of time to be spent on a particular ground school lesson, and also the approximate time to be devoted to the presentation of the elements of that lesson. For example, the time to be devoted to a ground school lesson on "maneuvering by reference to flight instruments" could be 90 minutes, with approximately the following time periods being used to present each of the elements: (a) straight-and-level flight—25 minutes, (b) turns—25 minutes, (c) climbs and descents—25 minutes, and (d) recovery from unusual attitudes—15 minutes.

An example of the approximate time to be devoted to the presentation and practice of the elements of a 90-minute instructional flight period on "ground reference maneuvers" could be: (a) preflight instruction—10 minutes, (b) instructor demonstrations—25 minutes, (c) student practice—45 minutes, and (d) postflight critique—10 minutes.

4. *Equipment.* This includes all instructional materials and training aids required to teach the lesson. For a ground school period, such items as films, slides, mockups, charts, computers, and reference materials should be included. For example, the equipment for a ground school period on "maneuvering by reference to flight instruments" could include the following: (a) an instrument panel mockup, (b) a copy of the FAA Instrument Flying Handbook, AC 61-27B, (c) selected slides on instrument flying, and (d) chalkboard and chalk.

For an instructional flight period on "ground reference maneuvers," the equipment should include at least: (a) a chalkboard for preflight discussion, (b) a copy of the FAA Flight Training Handbook, AC 61-21, and (c) an IFR visor for maneuvers reviewed.

5. *Instructor's actions.* This is a statement of the instructor's proposed procedures for presenting the elements of knowledge and performance involved in the lesson. Utilizing a combination of the lecture and the demonstration-performance methods, the instructor's actions during a ground school period on "maneuvering by reference to flight instruments" could be somewhat as follows: (a) discusses objective, (b) discusses concept of attitude instrument flying, (c) discusses and demonstrates straight-and-level flight from the standpoint of pitch, bank, power control, and trim, using an instrument panel mockup or chalkboard, (d) discusses and demonstrates turns from the standpoint of pitch, bank, power control, and trim, using an instrument panel mockup or chalkboard, (e) discusses and demonstrates climbs and descents from the standpoint of pitch, bank, power control, and trim, using an instrument panel mockup or chalkboard, (f) discusses and demonstrates recovery from unusual atti-

tudes, (g) assigns individual students the task of describing, and demonstrating, by means of an instrument panel mock-up or chalkboard, the control of an airplane by reference to flight instruments, and (h) critiques student presentation.

The instructor's action during an instructional flight period on "ground reference maneuvers" could be: (a) discusses objective, (b) diagrams "S" turns, eights along a road, and rectangular course on chalkboard, (c) demonstrates following a road and coaches student practice, (d) demonstrates "S" turns and coaches student practice, (e) demonstrates eights along a road and coaches student practice, (f) demonstrates rectangular course and coaches student practice, and (g) conducts postflight critique.

6. *Student's actions.* This is a statement of desired student responses to instruction. The student's actions during a ground school lesson on "maneuvering by reference to flight instruments" could be: (a) discusses objective, (b) listens, takes notes, and asks pertinent questions as the instructor lectures and demonstrates, (c) visualizes instrument maneuvers as the instructor lectures and demonstrates, (d) presents maneuvers, and (e) responds to questions posed by the instructor.

The student's actions during an instructional flight period on "ground reference maneuvers" could be: (a) discusses objective, (b) asks pertinent questions during preflight briefing, (c) at instructor's direction, reviews and practices power-off stalls and flight at minimum controllable airspeed, (d) performs ground reference maneuvers as directed by instructor, (e) asks pertinent questions both during flight and the postflight critique, and (f) responds to questions posed by the instructor.

7. *Completion standards.* This is the evaluation basis for determining how well the student has met the objective of the lesson in terms of knowledge and skill. For a ground school lesson on "maneuvering by reference to flight instruments," the evaluation may be accomplished by oral quizzing or by means of a short written test.

The evaluation at the end of an instructional flight period on "ground reference maneuvers" could be made from the standpoint of coordination, division of attention, orientation, proper wind drift correction, and accuracy in the maintenance of headings, altitude, and airspeed.

A sample lesson plan for a ground school period on "maneuvering by reference to flight instruments" appears on page 100. The "Instructor's Actions" are expanded and detailed on page 101. They may be further expanded as required by the individual instructor.

A sample lesson plan for an instructional flight period on "ground reference maneuvers" appears on page 102.

LESSON MANEUVERING BY REFERENCE TO FLIGHT INSTRUMENTS **STUDENT** ______ **DATE** ______

OBJECTIVE

- TO DEVELOP THE STUDENT'S UNDERSTANDING OF ATTITUDE INSTRUMENT FLYING AS RELATED TO STRAIGHT-AND-LEVEL FLIGHT, TURNS, CLIMBS AND DESCENTS, AND RECOVERY FROM UNUSUAL ATTITUDES

ELEMENTS

- STRAIGHT-AND-LEVEL FLIGHT
- TURNS
- CLIMBS AND DESCENTS
- RECOVERY FROM UNUSUAL ATTITUDES

SCHEDULE

- STRAIGHT-AND-LEVEL FLIGHT : 25
- TURNS : 25
- CLIMBS AND DESCENTS : 25
- RECOVERY FROM UNUSUAL ATTITUDES : 15

EQUIPMENT

- INSTRUMENT PANEL MOCKUP
- FAA INSTRUMENT FLYING HANDBOOK
- SELECTED SLIDES ON INSTRUMENT FLYING
- CHALKBOARD AND CHALK

INSTRUCTOR'S ACTIONS

- DISCUSS LESSON OBJECTIVE
- DISCUSS CONCEPT OF ATTITUDE INSTRUMENT FLYING
- DISCUSS, AND BY MEANS OF INSTRUMENT PANEL MOCKUP OR CHALKBOARD, DEMONSTRATE STRAIGHT-AND-LEVEL FLIGHT, TURNS, CLIMBS AND DESCENTS AND UNUSUAL ATTITUDE RECOVERIES
- ASSIGN INDIVIDUAL STUDENTS TASK OF DESCRIBING, AND DEMONSTRATING BY MEANS OF INSTRUMENT PANEL MOCKUP OR CHALKBOARD, THE CONTROL OF AN AIRPLANE BY REFERENCE TO FLIGHT INSTRUMENTS
- CRITIQUE STUDENT PRESENTATION AND ASK QUESTIONS

PRESENTATION

APPLICATION

REVIEW & EVALUATION

STUDENT'S ACTIONS

- DISCUSS LESSON OBJECTIVE
- LISTEN, TAKE NOTES, ASK PERTINENT QUESTIONS
- VISUALIZE INSTRUMENT MANEUVERS
- PRESENT MANEUVERS AND RESPOND TO INSTRUCTOR'S QUESTIONS

COMPLETION STANDARDS

- THE STUDENT SHOULD DEMONSTRATE, BY MEANS OF AN ORAL QUIZ OR WRITTEN TEST, THAT HE HAS AN UNDERSTANDING OF THE CONCEPT OF ATTITUDE INSTRUMENT FLYING AND OF THE PERFORMANCE OF BASIC FLIGHT MANEUVERS BY REFERENCE TO FLIGHT INSTRUMENTS

SAMPLE LESSON PLAN FOR A 90-MINUTE GROUND SCHOOL PERIOD

INSTRUCTIONAL AID	OUTLINE	NOTES
SLIDES ON INSTRUMENT FLYING	1. INTRODUCTION	1. ATTENTION--MAKE A STATEMENT OR ASK A QUESTION THAT RELATES LESSON TO STUDENT GOAL OF BECOMING A PROFICIENT INSTRUMENT PILOT. REVIEW PREVIOUS MATERIAL ON ATTITUDE INSTRUMENT FLYING AND GIVE TIE-IN BETWEEN THIS LESSON AND PREVIOUS LESSONS. 2. MOTIVATION--PROVIDE STUDENTS REASONS FOR NEEDING TO LEARN BASIC INSTRUMENT FLIGHT TECHNIQUE. 3. OVERVIEW--DISCUSS LESSON OBJECTIVE AND KEY IDEAS TO BE PRESENTED.
INSTRUMENT PANEL MOCKUP CHALKBOARD SLIDES ON INSTRUMENT FLYING FAA INSTRUMENT FLYING HANDBOOK	2. DEVELOPMENT	1. DISCUSS CONCEPT OF ATTITUDE INSTRUMENT FLYING. 2. PRESENT STRAIGHT-AND-LEVEL FLIGHT ON MOCKUP FROM STANDPOINT OF PITCH, BANK, POWER, AND TRIM CONTROL. 3. PRESENT TURNS ON MOCKUP FROM STANDPOINT OF PITCH, BANK, POWER, AND TRIM CONTROL. 4. PRESENT CLIMBS AND DESCENTS ON MOCKUP FROM STANDPOINT OF PITCH, BANK, POWER, AND TRIM CONTROL. 5. PRESENT RECOVERY FROM UNUSUAL ATTITUDES ON MOCKUP. 6. ASSIGN INDIVIDUAL STUDENTS TO PRESENT INSTRUMENT MANEUVERS ON MOCKUP MONITOR STUDENT PRESENTATION AND MAKE APPROPRIATE COMMENTS.
FAA INSTRUMENT FLYING HANDBOOK	3. CONCLUSION	1. RETRACE IMPORTANT POINTS RELATED TO ELEMENTS OF KNOWLEDGE PRESENTED AND RELATE THEM TO THE LESSON OBJECTIVE. 2. DETERMINE WHETHER OR NOT STUDENTS HAVE MET OBJECTIVE OF LESSON BY SHORT ORAL QUIZ OR WRITTEN TEST. 3. ASSIGN STUDENTS TO STUDY CHAPTER V OF THE FAA INSTRUMENT FLYING HANDBOOK AS IT RELATES TO MAGNETIC COMPASS, TURNS TO PREDETERMINED HEADINGS AND TIMED TURNS. GIVE TIE-IN BETWEEN THIS LESSON AND NEXT LESSON.

"INSTRUCTOR'S ACTIONS" EXPANDED AND DETAILED

PRESENTATION + REVIEW & EVALUATION

LESSON GROUND REFERENCE MANEUVERS **STUDENT** ____________ **DATE** ________

OBJECTIVE

- TO DEVELOP THE STUDENT'S SKILL IN PLANNING AND FOLLOWING A PATTERN OVER THE GROUND COMPENSATING FOR WIND DRIFT AT VARYING ANGLES

ELEMENTS

- USE OF GROUND REFERENCES TO CONTROL PATH
- OBSERVATION AND CONTROL OF WIND EFFECT
- CONTROL OF AIRPLANE ATTITUDE, ALTITUDE, AND HEADING

SCHEDULE

- PREFLIGHT DISCUSSION :10
- INSTRUCTOR DEMONSTRATIONS :25 } Presentation
- STUDENT PRACTICE :45 Application
- POSTFLIGHT CRITIQUE :10 Review & Evaluation

EQUIPMENT

- CHALKBOARD FOR PREFLIGHT DISCUSSION
- IFR VISOR FOR MANEUVERS REVIEWED

INSTRUCTOR'S ACTIONS

- PREFLIGHT-DISCUSS LESSON OBJECTIVE. DIAGRAM "S" TURNS, EIGHTS ALONG A ROAD, AND RECTANGULAR COURSE ON CHALKBOARD
- INFLIGHT-DEMONSTRATE ELEMENTS. DEMONSTRATE FOLLOWING A ROAD, "S" TURNS, EIGHTS ALONG A ROAD, AND RECTANGULAR COURSE. COACH STUDENT PRACTICE

Presentation

Application

- POSTFLIGHT-CRITIQUE STUDENT PERFORMANCE AND MAKE STUDY ASSIGNMENT

Review & Eval.

STUDENT'S ACTIONS

- PREFLIGHT-DISCUSS LESSON OBJECTIVE AND RESOLVE QUESTIONS
- INFLIGHT-REVIEW PREVIOUS MANEUVERS INCLUDING POWER-OFF STALLS AND FLIGHT AT MINIMUM CONTROLLABLE AIRSPEED. PERFORM EACH NEW MANEUVER AS DIRECTED.
- POSTFLIGHT-ASK PERTINENT QUESTIONS

COMPLETION STANDARDS

- STUDENT SHOULD DEMONSTRATE COMPETENCY IN MAINTAINING ORIENTATION, AIRSPEED WITHIN 10 KNOTS, ALTITUDE WITHIN 100 FEET, AND HEADINGS WITHIN 10 DEGREES, AND IN MAKING PROPER CORRECTION FOR WIND DRIFT.

- PLANNING
- COORDINATION
- ORIENTATION
- DIVISION of ATT.
- CONTROL w/ WIND

SAMPLE LESSON PLAN FOR A 90-MINUTE INSTRUCTIONAL FLIGHT PERIOD

✓ 2-12-91

SECTION THREE
THE AIRCRAFT MAINTENANCE INSTRUCTOR

CHAPTER XII. AIRCRAFT MAINTENANCE INSTRUCTOR CHARACTERISTICS AND RESPONSIBILITIES

Professional aircraft maintenance instructors should, above all, be highly experienced and technically competent in the areas in which they plan to teach. Along with technical competence, the ability to convey knowledge and skill to others is essential. In addition, the aircraft maintenance instructor must have qualities which are universally associated with professionalism.

CHARACTERISTICS OF A PROFESSIONAL AIRCRAFT MAINTENANCE INSTRUCTOR

Good Grooming

Since professional aircraft maintenance instructors are in positions of responsibility and authority, it is essential that they preserve and enhance the image which students have the right to expect of them.

To present a proper image, professional instructors should wear clothing that is appropriate to the environment in which training is being conducted. Obviously, students do not expect their instructor to be dressed in a business suit while giving shop instruction, but they do have a right to expect the instructor to be neat, clean, and appropriately dressed. Appearing before a class in dirty, ragged coveralls immediately destroys the instructor's professional image.

Attitude

Aircraft maintenance instructors should indicate, through both word and action, that they hold their profession in high esteem. Such a positive attitude will invariably be adopted by the students. Conversely, a careless and perfunctory performance on the job will immediately result in a loss of student respect and motivation.

Language

An aircraft maintenance instructor should use language which is appropriate to the material being taught and the educational level of the students. A skilled instructor will present difficult material in the most simple and understandable terms possible, without giving the impression of "talking down" to the students. Public speaking clubs or college courses are often helpful in developing and improving an instructor's ability to give oral presentations.

ROLE PLAYING BY THE PROFESSIONAL AIRCRAFT MAINTENANCE INSTRUCTOR

The Instructor as a Practical Psychologist

Aircraft maintenance instructors are required to play a variety of roles. One role is that of practical psychologist. In this role, the instructor must accurately analyze and evaluate student attitudes and mental abilities in order to present material in such a way that maximum learning is achieved by the group in a given time period. Instructors should attempt to determine the rate at which the students are able to absorb instruction and adjust their teaching to the average learning rate of a particular group. In any group, instructors must deal with the problem of slow and fast learners. An instructor must expect to spend additional time and effort with a slow student to assure an acceptable level of knowledge and skill. A fast learner should be given additional assignments or projects to prevent boredom. Regardless of whether a student is a fast or slow learner, the instructor's assignments should challenge ability to the limit. Instructors must use imagination and exercise constant effort in order to stimulate and maintain a high level of interest among the variety of students that compose a typical aircraft maintenance class.

The instructor must be aware of any student physical defects that may hamper learning; e.g., defective sight or hearing. This determination can best be made through interviews, personal observation, and the examination of student records. Students with correctable physical deficiencies should be encouraged to obtain professional advice and assistance prior to starting a long, difficult training program, because the training may place additional strain on a weakened faculty. Obviously, students with defective sight or hearing should be placed near the front of an instructional area. An instructor should realize that left-handed students are not handicapped but have a physical characteristic that should be taken into consideration during manipulative skill demonstrations. An alert, professional instructor will readily recognize the individual handicaps and traits of students and will make every effort to see that they are as comfortable and effective as the classroom or shop environment will allow.

Various intelligence and aptitude tests are available to aircraft maintenance instructors; however, these are only indicators and may be misleading with regard to actual performance in the shop or classroom. A student's physical condition or personal problems may prevent high scoring on some tests. Most students taking aircraft maintenance training are highly motivated, and this may compensate for some deficiencies revealed by tests.

The Instructor as an Educator

An aircraft maintenance instructor plays the role of educator in planning and guiding classes. This planning and guidance includes, but is not limited to, the following: (1) adequately organizing lectures, (2) guiding discussion groups, and (3) planning shop projects. Instructors may find their effectiveness increased during discussion periods and shop sessions if slow and fast learners are placed in separate groups.

In organizing the material to be presented during an aircraft maintenance course, the instructor will find it necessary to select and evaluate topics to be included. Supportive subjects should be placed near each other and horizontal as well as vertical learning built into the instruction. Subjects that reinforce one another enhance learning. In developing activity plans, the instructor must integrate theoretical knowledge presented during classroom instruction with skills to be learned during shop periods.

The Instructor as a Judge

An aircraft maintenance instructor plays the role of judge in evaluating student knowledge and skill. It is important that this evaluation be as fair and comprehensive as possible. This can be accomplished through oral quizzing, written tests, and observing student performance during shop periods. A competent instructor will use all three of these evaluation methods to locate deficiencies in student learning, to motivate achievement, and to determine weak points of instruction.

Oral questions submitted to a student should be clear, concise, and appropriate to the required knowledge level at a particular stage of training. Written tests should be constructed following the guidelines set forth in Chapter VII. An analysis of written test results appears in the Appendix. While in the role of judge of student performance, the instructor must resist any tendency to downgrade a student because of personal prejudice. Such things as hair style, clothing, and mannerisms have no bearing on a student's ability to acquire essential knowledge and skill. A professional instructor is unbiased and completely objective.

TECHNOLOGICAL ADVANCES

The professional aircraft maintenance instructor must stay current with regard to the latest technological advances in the aviation industry in order to assure the training of competent students with salable skills. The instructor can observe other schools in the aircraft maintenance field to determine if there is any marked difference in the material being taught. Various trade publications provide information on new materials, equipment, and aircraft. The instructor should, if possible, take students on field trips to local aircraft and instrument factories. Attendance at aviation seminars and discussions with members of professional aviation organizations provide important personal contacts of value in staying abreast of the industry. Also, there are a number of government publications available which will assist the instructor in this area.

SAFETY

The aircraft maintenance instructor has a definite responsibility for planning a lesson on safety in the classroom and the shop. The safety lesson should be presented to a class at the beginning of training and should be followed by an oral quiz or written test.

It is becoming increasingly the school's responsibility to provide students with such items as safety glasses, hard hats, and safety shoes. If the school does not provide them, it is the instructor's responsibility to request that students wear such protective equipment. The instructor is also responsible for determining that "warning" and "caution" signs are suitably placed in the shop area, and for enforcing established safety rules.

EDUCATIONAL PHILOSOPHY

The professional aircraft maintenance instructor should know the accepted philosophies of industrial and vocational education. The instructor's goal should be to develop students as competent technicians in the field of aircraft maintenance and also to motivate them to advance to other aviation-related fields. To clarify the instructor's thinking in this area, it is recommended that a statement of educational philosophy, as it applies to aircraft maintenance training, be developed. Over a period of time, and as a result of experience, the instructor may find it necessary to modify and improve this statement.

CHAPTER XIII. INTEGRATED JOB TRAINING

Integrated job training, as it applies to aircraft maintenance instruction, is the combining of knowledge teaching with shop activities. The integration of theoretical knowledge with manual skill is a teaching method with almost unlimited possibilities. It lends itself well to group projects and activities, thereby creating interest and enhancing motivation. Instructors must set worthy goals which are within the students' ability to reach, if motivation is to be maintained at a high level. The evident final student goal, of course, is the attainment of competence as an aircraft mechanic and the development of a salable skill. Another important student goal is the development of a professional sense of pride. Without this positive attitude, the high status that certificated aviation mechanics presently enjoy cannot be maintained.

STUDENT GOALS TO BE ATTAINED

Professional Competence

The aircraft maintenance instructor bears a heavy responsibility for producing airmen who are thoroughly trained in the basic fundamentals of aircraft maintenance. Students must assimilate these basics in the training environment of the classroom and shop if they are to perform with competence on the job. The instructor can encourage the development of competence, pride in workmanship, and professionalism through the grading of projects. Grades should be reduced for lower-than-standard-quality work and poor practices. Shortcuts and makeshift repairs should definitely be discouraged.

Salable Skill

The development of a salable skill is an important student goal. Failure to attain this objective reflects adversely on the instructor's teaching ability. If students have not achieved sufficient expertise while in school to enable them to acquire and hold paying jobs as aircraft mechanics, the training has been in vain, and the instructor has clearly failed. In this connection, it should be noted that training should be conducted using up-to-date equipment. A graduate cannot expect to attain a salable skill if only obsolete engines, airframes, and components are used by the school.

Pride in Work and Profession

Aircraft maintenance is a field of endeavor in which journeymen still take pride in turning out quality work. It is an important instructor responsibility to see that this tradition of excellence, which is characteristic of skilled professional mechanics, is passed on to students. The instructor's own performance should be an example to the students. The instructor may reinforce pride in work through individual commendation of students who have done outstanding work.

Positive Attitude

A positive attitude is an asset in any field of endeavor, including aircraft maintenance. It is a personal characteristic that reinforces itself. An individual with a positive attitude thinks success can be achieved, and when it is actually realized, the individual's attitude and thinking are reinforced. An aircraft maintenance instructor can encourage a positive attitude on the part of students by means of grades, recognition, and commendations. The development of this attitude can also be facilitated by planning projects that give students "success experiences."

The FAA Aviation Mechanic Certificate

The award of the FAA Aviation Mechanic Certificate is an obvious student goal. In a sense, however, the certificate is not the real goal. It is only a formal recognition by the FAA that the student has attained the knowledge and skill required by regulations. Knowledge and skill are the real goals. With the awarding of the certificate, the student is ready to assume the demanding responsibilities, and also to enjoy the satisfactions, that are associated with the profession.

TEACHING INTEGRATED KNOWLEDGE AND SKILL

The aircraft maintenance instructor is responsible for planning student activities so that theory and skill are taught in a unified manner. The integrated approach should be used in the two basic areas of aircraft maintenance training, airframe and powerplant. The following techniques and procedures will be of value to the instructor in accomplishing the objective of integrated training.

Lecture

During the first few lessons, the instructor should use the lecture method to acquaint students with the plans, goals, and activities of the course. As the course progresses, the lecture should be combined with other teaching methods; i.e., guided discussion and demonstration-performance.

Group Activities

The instructor may divide a class into groups for the purpose of planning, discussion, or the accomplishment of projects. This procedure involves delegating authority to the group and permitting the members to participate in their own learning. The instructor makes an assignment, then holds the group responsible for planning and completing the project. The group should complete all planning before beginning work. It should also prepare demonstrations and oral reports to be given to the rest of the class. Each group member should fully participate in the work and turn in a written report of the activities that were involved in the accomplishment of the project.

Demonstrations

Group demonstrations provide an excellent way of integrating the teaching of knowledge and skill. Demonstrations should include any instructional aid that makes the learning activity more easily understood. Posters, sequence charts, chalkboard drawings, and models are all effective aids which the instructor should use when appropriate to a demonstration. After a demonstration, each student should turn in a written report describing what has been covered.

Written Reports

Written reports, such as those mentioned in the preceding paragraphs, can be used advantageously during integrated instruction. The experience of preparing a report helps both strong and weak students retain instruction. If a report is prepared by a group, the more proficient students are aided by assisting those who are less competent in a given area. The weaker students benefit from the assistance of the stronger students. Students who are weak in a certain area; e.g., fuel systems, may be assigned the task of researching several types of systems, then submitting a report on them. The instructor may specify the points to be emphasized in the report, for example, the advantages and disadvantages of each system. When a report is preceded by planning and thought, it has the potential of teaching students through their own efforts. A student who researches an area of knowledge and prepares a report, goes through a thinking process and answers many questions. This procedure usually results in the achievement of permanent learning.

Question Sessions

Question sessions may be used to stimulate student interest in a specific knowledge or skill area. This activity requires careful instructor preparation and the establishment of a friendly, relaxed atmosphere. The instructor should give each student an opportunity to answer several questions and should never intentionally embarrass a student. A question session may be used to evaluate both student progress and instructor effectiveness. Also, specific student problems and misunderstandings can be identified at this time. A more complete discussion of oral questioning is given in Chapter VII.

Field Trips

Since field trips give students the opportunity to acquire knowledge and observe shop work simultaneously, they should, if possible, be included in the learning activities of each aircraft maintenance class. The aircraft industry to be visited should be contacted ahead of time for a plant tour. The instructor should require each student to prepare an outline of the learning to be expected. Also, the students should list the processes and procedures to look for during the tour. In addition to technical questions about plant activities, the students may ask about such things as job availability, seasonability of work, professional practices on the job, salaries, and company benefits. After the completion of the tour, each student should write a report on the activities observed and knowledge gained. If the instructor uses this approach, the students will derive many benefits from their observation of the real work in the aviation industry.

Theory and Shop

For effective learning, the aircraft maintenance instructor should combine the teaching of theory with shop projects to the greatest extent possible. It is important that shop projects be planned to parallel theory being taught, since these reinforce one another. The shop projects should be accompanied with as much theory as can be efficiently utilized. A film of a complete powerplant overhaul shown before powerplant shop periods may have value; however, a set of slides on valve maintenance shown to coincide with a shop project on valve repair would be more effective. When the instructor is teaching hydraulic systems, a mockup of a system should be in front of the class. This will enable the students to immediately test the theory being taught. Also, the instructor can insert defects in the system, thereby giving the students practice in troubleshooting. An instructor who combines theory with actual shop performance will find that teaching is easier and more effective.

Student Evaluation

The evaluation of student achievement may be accomplished by one or more of the methods described in Chapter VII. Of course, the evaluation method used should be appropriate to the area being tested.

Abstract knowledge, or theory, can best be evaluated through oral quizzing or written tests. The instructor may use these two evaluation methods to measure student achievement during a lesson, during a particular stage of training, or at the end of a course.

Manual skills are evaluated by the instructor through a direct observation of student performance. Both the instructor and student should be fully aware of the qualities and standards of performance required for a certain grade at a particular stage of training. In evaluating a project, the instructor should outline the areas to be graded; i.e., quality, quantity, workmanshop, planning, and time required for completion. The evaluation of skills should be fair and firm, but also accomplished in a manner that will motivate the student to higher achievement.

Tests are of value not only in revealing student weaknesses and providing a basis for grades, but also in indicating the effectiveness of instruction. Tests can also point out teaching areas that need greater emphasis or perhaps a different approach.

Tests, in addition to being evaluation instruments, have either a positive or a negative motivating effect. Ideally, they should motivate a student positively, reinforce thinking, and stimulate greater effort. The student should believe success is possible.

Progress charts should be prepared on the basis of test results. These charts enable both the instructor and the student to visually observe progress. They also provide a means of recognizing student achievement and of improving motivation.

Instructional Aids

Instructional aids have been discussed in considerable detail in Chapter VIII. The aircraft maintenance instructor will find such aids of great value if they are skillfully used. To gain maximum benefit from an aid, the instructor should first determine which aid is to be used during a particular lesson and also become familiar with the aid's advantages and disadvantages. The planning of an activity that utilizes an instructional aid should include an alternative plan to be followed in the event the aid is not available or a failure occurs. Aids should not be used so extensively that the instructor's teaching skill deteriorates. Aids should be carefully selected and prepared, and should be suited to the material being taught. Several useful instructional aids are discussed below.

Overhead Projector. This aid is useful because it can be used to prepare for an activity ahead of time. Also, the material on overhead projector slides can easily be changed, thus allowing for updating and the correction of errors.

Movies and Slides. These aids are effective when used to parallel a lesson. Preplanning is important here, since the instructor must locate, preview, and have available movies and slide sets that are appropriate to the lesson being given. A side benefit of movies is that they give students a break in the routine of regular class work.

Mockups. Powerplants and other aircraft components with cutaway viewing areas are instructional aids which can be used to good advantage. This type of aid can easily be used to integrate theory and shop. They add realism to teaching, in that students are able to see the internal parts and observe the operation of a piece of equipment used in modern aircraft.

Pictures, Posters, Charts, and Chalkboards. Instructional aids of this type are more economical to obtain and use than mockups of actual equipment. They can be constructed in the classroom or shop as individual or group projects. Detailed charts and diagrams should be prepared well in advance of the time they are needed in class. When they are not in use, it is advisable to place them in a storage area so they will not distract the attention of the students. The chalkboard is the most common instructional aid and is used by the instructor during almost every class period. Chapter VIII includes a list of recommended practices concerning the use of this aid, which the instructor should review. Some instructors fall into the habit of attempting to lecture or explain a point while facing the chalkboard. This is a poor practice and should be avoided.

Bulletin Board. The entire class should participate in posting material on the bulletin board. The instructor should motivate interest in the bulletin board, but the students should be responsible for its contents. The instructor may encourage students to read the bulletin board by posting a few questions which relate to the material currently being covered in class. Also, material on the latest innovations in the aircraft maintenance field should be posted and their relationship to current lessons shown.

Instructor Qualifications

To adequately conduct integrated job training, the aircraft maintenance instructor should, if possible, have formal training in educational psychology, teaching methodology, and test and measurement techniques. If formal training is not feasible, a thorough study of the contents of this handbook will provide the instructor with adequate knowledge of how students learn, the various teaching methods and techniques, and the ways learning can be evaluated. It is highly desirable that the instructor be a certificated FAA Airframe and Powerplant Mechanic with extensive practical experience.

CHAPTER XIV. PLANNING AIRCRAFT MAINTENANCE INSTRUCTIONAL ACTIVITIES

Both prior to and during an aircraft maintenance training course, the instructor will find it necessary to do a considerable amount of planning. Planning must take place if course objectives are to be successfully accomplished. Areas in which planning is essential are the: (1) acquisition of new training material, (2) integration of new material into the training course, (3) utilization of new material, (4) discarding of obsolete material, and (5) planning of lessons.

ACQUISITION OF NEW TRAINING MATERIAL

Aircraft Manufacturers

Aircraft manufacturers, subcontractors, and overhaul companies are excellent sources of new and up-to-date material for use by aircraft maintenance schools. The instructor should contact these sources frequently for current procedures and information that can be utilized in airframe and powerplant course curricula.

Aircraft Maintenance Schools

An aircraft maintenance instructor should maintain a close liaison with successful airframe and powerplant schools and communicate with them at frequent intervals to determine their sources of information and materials. An exchange of views on the problem of keeping school materials current will also be helpful.

Professional Organizations and Seminars

There are a number of professional organizations from which an instructor can derive benefit through the exchange of ideas and the discussion of problems related to the acquisition of training materials and their utilization in a training program. Through discussions with people in the same field, the instructor may also find solutions to various shop and classroom problems. In addition, such discussions may aid the instructor in maintaining an up-to-date curriculum.

Aviation-related seminars are valuable sources of information. A seminar is a virtual marketplace for locating new ideas for doing old jobs. The latest systems, procedures, and data are often displayed at a seminar. Also, seminars provide an opportunity for the instructor to meet and talk with other members of the aviation industry.

Magazines, Professional Journals, Technical Books, and Texts

Magazines and professional journals are excellent sources of information on the latest equipment and training material. The instructor should investigate this area and make several such periodicals available to the class for supplemental reading.

New technical books and texts are also valuable sources of training material and ideas. The instructor should build a technical library that includes a variety of aviation-related maintenance technical books.

These are just a few of the sources of new training material. The competent professional instructor will use imagination and ingenuity in locating new material and equipment and integrating it into the training course.

INTEGRATING NEW MATERIAL INTO TRAINING ACTIVITIES

Evaluating New Material

The instructor must evaluate new training material for usability. Also, the material should be examined closely to determine that it does not duplicate existing training material and that it will be of value in helping students attain the course objectives. To justify its use in a training program, the new material must aid in developing students into competent aviation mechanics and assist them in passing the FAA airframe and powerplant tests.

Utilizing New and Discarding Obsolete Material

With forethought and planning, the instructor can work new material into a training program in a smooth and nondisruptive manner. The new material should be integrated into lessons to further the accomplishment of particular learning outcomes. With increasing experience, the instructor will become more and more adept in this area of instructional planning.

Before discarding an old item of training material, the instructor should be certain that it is no longer useful in attaining course objectives and that more up-to-date material is available to replace it. The instructor should attempt to anticipate the need for replacing obsolete training materials and try to introduce the new materials in an orderly manner.

Lesson planning has been covered in some depth in Chapter XI. The planning and construction of organized lesson plans is an important area of responsibility to which the aircraft maintenance instructor should devote considerable time and effort. A desk top index of lesson plans should be developed. This aids the instructor in organizing the activities for each classroom or shop period, helps in arranging the sequence of instructional units, and tells the students what areas are to be covered during the course. Also, the students are provided information on additional support material to study. The lesson plan gives order and direction to the activities of a particular class period. A lesson plan may be long and detailed or it may be brief. In any case, it should be thought out and developed to the extent that instructional activities progress smoothly and efficiently.

APPENDIX

CONTENTS

HOW TO OBTAIN REFERENCE AND TRAINING MATERIALS

Reference and training materials, such as Federal Aviation Regulations and advisory circulars, may be ordered from the address given below. For your convenience in ordering publications sold by the Superintendent of Documents, an order form is included in the back of this handbook.

Superintendent of Documents
U.S. Government Printing Office
Washington, D.C. 20402

To obtain the latest information regarding Federal Aviation Regulation prices, number of changes, and ordering information, send for a free copy of "Advisory Circular 00–44, Status of Federal Aviation Regulations" from the address given below. Free FAA publications may also be obtained from this office:

U.S. Department of Transportation
Publications Section, TAD–443.1
Washington, D.C. 20590.

If you are presently on one of FAA's advisory circular mail lists, you will automatically receive Advisory Circular 00–44. If not, and you wish to be placed on the mail list to receive revised copies as issued, send your name and address to:

U.S. Department of Transportation
Distribution Requirements Section, TAD–482.3
Washington, D.C. 20590.

To receive the latest information on how to obtain the Airman's Information Manual, free advisory circulars, and other FAA material, consult the latest Advisory Circular Checklist, AC 00–2. A copy of this checklist may be obtained free of charge by sending your request to:

U.S. Department of Transportation
Publications Section, TAD–443.1
Washington, D.C. 20590.

Exam-O-Grams may be obtained free of charge (one copy per request), and names may be added to the mailing list by writing to:

Department of Transportation
FAA Aeronautical Center
Flight Standards National Field Office
Examinations Branch, AFS–590
P.O. Box 25082
Oklahoma City, Oklahoma 73125.

NOTICES OF PROPOSED RULE MAKING (NPRMs)

The Government Printing Office will send NPRMs to all purchasers of the Federal Aviation Regulation containing the particular regulation concerned.

FILMS

FAA—Order free catalog from:

FAA Film Library, AAC–44.5
FAA Aeronautical Center
P.O. Box 25082
Oklahoma City, Oklahoma 73125.

USAF—For information write:

USAF Central Audio-Visual Library
AF Audio-Visual Center
Norton AFB, California 92409.

A number of excellent films are also available to instructors from industry sources.

ANALYSIS OF AIRCRAFT MAINTENANCE AIRMEN WRITTEN TEST RESULTS

* * * * * *

The following description of an analysis of written test results is presented for use by the aircraft maintenance instructor who has a working knowledge of elementary educational statistics. The instructor who does not have such knowledge may acquire it through study of any standard text on educational tests and measurements.

* * * * * *

To improve professional ability and potential, the aircraft maintenance instructor should be able to analyze written test results. The following is an example using the raw test scores of sixteen students on a given test for analysis. This example is used to describe the terms and explain their benefit to the instructor. An analysis of raw test scores achieved by students on a written test appears on page 120. The test scores are divided into score intervals of three test scores per interval. This allows the instructor to compress the sixteen raw scores into ten groups of scores. The median, which separates the top half of the group from the bottom half, in this example, is 80.5, the center of the raw test scores. This serves as a reference for analyzing the score distribution.

This reference can be used by the instructor for comparison purposes. For example, a comparison of the arithmetic mean with the median will show the instructor if the students are performing above, on, or below the median. Students having difficulty on a written test are easily identified. This also points to the tests and test items which should be examined for defects. An arithmetic mean skewed above the median, may indicate that the test is too easy. Also, this may indicate that the instructor has been highly effective in teaching the class. When both the tests and the instruction have been modified several times, based on this type of analysis, the arithmetic mean should be very close to the median, as in the test analysis example which follows.

The semi-interquartile range (Q) is half of the middle 50 percent. In the example, this is 4. This represents one-fourth of the raw scores. It also represents the average deviation from the median to either quartile (the 25 percent or 75 percent percentile).

Standard Deviation (SD) is another means of determining the spread or distribution of a group of scores in relation to the arithmetic mean. The SD in this example is 8.16 and represents the average or spread of the deviations from the arithmetic mean. Also, SD is a yardstick for evaluating student performance. Both the arithmetic mean and SD are valuable tools for use in interpreting test scores and comparing the test score spread of different classes. A student test score above the median may represent a high score for a particular class; however, when compared with ten preceding classes, the score may be very low. By comparing the test results of different classes, the instructor can eliminate test items that are not performing adequately. The instructor may also find that the test items are acceptable and simply more instructional time needs to be applied in a specific area.

The analysis of written test results has the potential of improving confidence in the student training results achieved by aircraft maintenance schools and instructors.

Raw Test Scores	*Score Interval*	*Tallies*	*Frequency*	X'	FX'	$F(X')^2$
95	93–95	1	1	5	5	25
91	90–92	1	1	4	4	16
88	87–89	1	1	3	3	9
87						
86	84–86	11	2	2	4	8
85						
83						
82	81–83	111	3	1	3	3
81					+19	
80						
79	78–80	111	3	0		
78						
76	75–77	11	2	−1	−2	2
73	72–74	1	1	−2	−2	4
71	69–71	1	1	−3	−3	9
67	66–68	1	1	−4	−4	16
		16			−11	
					+ 8	122

The median and arithmetic mean differ only if scores are skewed or piled up on one side.

Median = 80.5

$$\text{Mean} = \frac{(\text{Sum of } FX^1)}{N} (\text{Interval}) + \text{Arbitrary Origin}$$

$$\text{Mean} = \frac{(+8)}{16} (3) + 79$$

$$\text{Mean} = .5\ (3) + 79$$

$$\text{Mean} = 1.5 + 79$$

$$\text{Mean} = 80.5$$

Standard deviation

$$\text{S.D.} = 3\sqrt{\frac{122}{16} - \left(\frac{8}{16}\right)^2} = 8.16$$

16 total test scores

$$\frac{50\%}{8.00 \text{ middle } 50\%}$$

$$\text{Semi-interquartile range } Q = \frac{25\% - 75\%}{2}$$

$$Q = \frac{4. - 12.}{2}$$

$$Q = 4.$$

INDEX